Introduction to HIDA DISTRIBUTIONS

Introduction to HIDA DISTRIBUTIONS

Si Si

Aichi Prefectural University, Japan

NEW JERSEY • LONDON • SINGAPORE • BEIJING • SHANGHAI • HONG KONG • TAIPEI • CHENNAI

Published by

World Scientific Publishing Co. Pte. Ltd.
5 Toh Tuck Link, Singapore 596224
USA office: 27 Warren Street, Suite 401-402, Hackensack, NJ 07601
UK office: 57 Shelton Street, Covent Garden, London WC2H 9HE

British Library Cataloguing-in-Publication Data
A catalogue record for this book is available from the British Library.

INTRODUCTION TO HIDA DISTRIBUTIONS

ISBN-13 978-981-283-688-5
ISBN-10 981-283-688-8

Printed in Singapore.

To the memory of my parents U Khin Aung - Daw Su Su

Preface

We first recall the Pioneers' work in the area of Stochastic Analysis. N. Wiener wrote a paper "Differential space" in 1923. Actually he studied a Brownian motion and its functionals. Then he wrote "Homogeneous chaos" 1938. Then, K. Itô extended his work, by introducing the notion of "Multiple Wiener integrals" in 1951. Following these works, the researchers have been working on the analysis of functionals of Brownian motion. To this direction P. Lévy's work on functional analysis, in particular, analysis on function space, is a great contribution.

In 1975, Takeyuki Hida proposed a new stochastic analysis in "Lectures on Brownian functionals", Carleton University Lecture Notes. He introduced a "generalized" Brownian functionals which are nowadays called "Hida distributions" or "generalized white noise functionals". L. Streit firstly named the subject as "Hida distributions".

We recall the history of classical analysis. In 1950, L. Schwartz introduced a space of "Distributions" which are generalized functions (not random).

T. Hida succeeded in the case of random functional, indeed Brownian functionals, in the very advanced level. After that, I. Kubo and S. Takenaka have given another nice setup in 1980. Together with contribution by many others, they are now called white noise functionals and they have developed to Hida distributions. In order to establish the calculus of those generalized white noise functionals, it is necessary to introduce tools of the analysis, differential operators and integration, and further Laplacians. This direction has extensively developed in these years.

White noise analysis has good applications in quantum physics; in particular in Feynman path integrals. The original paper is written by T. Hida and L. Streit in 1983. Significant developments of Feynman path integrals in their approach have been done by researchers in Germany, Philippines and Italy.

We can also see interesting applications in quantum mechanics. These days, applications to Biology and Economics became very popular.

White Noise Analysis has two significant advantages.

1. One is the space of Hida distributions (generalized white noise functionals).
2. Infinite dimensional rotation group with a great contributions.

To explain the space of generalized white noise functionals, we start with white noise, the time derivative of a Brownian motion $B(t)$ denoted by $\dot{B}(t)$.

As is known $B(t)$ is continuous but not differentiable, so time derivative does not exist in the ordinary sense. On the other hand, white noise is atomic infinitesimal random variable. T. Hida wishes to use $\dot{B}(t)$ itself to express random events that are evolutional as the time t runs. He used the trick as follows.

i) Smeared variable $\int \dot{B}(t)f(t)dt$ is well-defined in the classical sense, where $f(t)$ is a smooth function.
ii) From classical to modern theory (that is white noise theory): he takes the delta function δ_t in place of f. Then, smeared variable becomes $\dot{B}(t)$. To be able to use δ_t, it is required to utilize the Sobolev space of order -1 which contains the delta function. Using the delta function ($f(t)$ in the above integral is changed to δ_t), $\dot{B}(t)$ can be defined.
iii) Next he made a development to nonlinear functionals of white noise. Basic functions of the variables (i.e. white noise) should be polynomials. Unfortunately, powers of white noise cannot be well defined. Therefore "renormalizations" of the monomials is necessary. In fact they are Hermite polynomials of white noise. Note that this method cannot be seen in classical stochastic analysis.
iv) Collection of those renormalized polynomials span the space of generalized white noise functionals, denoted by $(L^2)^-$ which is quite useful.
v) The infinite dimensional rotation group, which has been introduced by H. Yoshizawa, characterizes the white noise measure, the probability distribution of $\{\dot{B}(t), t \in R^1\}$. White noise analysis has an aspect of the harmonic analysis arising from the rotation group.

I hope that this monograph will help to understand better the white noise analysis and to its further developments.

This monograph is written based on the ideas learned from several lectures given by Professor Takeyuki Hida and the collaboration with him for many years. I would like to express my deep gratitude to Professor Takeyuki Hida for his encouragement and valuable suggestions.

I would also like to express my gratitude to all of my teachers in Myanmar, especially to Professor Dr. Chit Swe, my supervisor at Rangoon University and Professor U Sein Min.

In addition, I would like to thank all the members of white noise (probability) seminar in Nagoya.

Si Si
Faculty of Information Science
Aichi Prefectural University
Aichi-Ken, Japan
July 2011

Contents

Chapter 1

Preliminaries and Discrete Parameter White Noise

1.1 Preliminaries

Our aim is to provide an introduction to the white noise analysis, the so-called Hida distribution theory, which has extensively developed these years. For this purpose we shall remind the basic idea of the analysis and explain how the theory has been setup by T. Hida.

The theory was started with the study of random complex systems which are assumed to be parametrized by the time and/or space and to be developed as the time goes by. This monograph will follow Hida's original idea.

T. Hida established the white noise analysis in line with the following steps:

Reduction, Synthesis, Analysis.

First, we give a quick explanation what each step means and how they are proceeded.

Reduction means the following. A given random complex system may be expressed mathematically as a system of many random variables which are mutually correlated to each other. It would be easier to deal with them in probability theory if they were mutually independent. If so, the probability distribution of the total system is just the direct product of the distributions of those variables. We are now familiar with discussions on the product measure.

In general, such an independence cannot always be expected. We are, therefore, asked to make the system to be that of mutually independent variables. Such independent system should have the same information (which can be described in terms of sigma-fields) as the original one. It is the best if this can be done, but it is not easy to do so, in general. Hence we are given a hard problem. Another serious problem arises. Namely, if the system involves finite or countably infinite independent variables, then we can proceed in the established manner, while if it involves continuously many variables, then we have to find a method to let the family be separable, that is, can be analyzed in the ordinary method.

Synthesis. Once a desired system of independent random variables, denoted by $\mathbf{Y} = \{Y_\alpha, \alpha \in A\}$, is obtained, we shall proceed to the problem how to determine the function $f(Y_\alpha, \alpha \in A)$ that describes the given random complex system. When the number of the elements Y_α is uncountable, the function f cannot be arbitrary chosen, as we shall see in Chapter 3. We claim that at least elementary functions, like polynomials, should be accepted with suitable modifications if necessary.

Analysis. After the variables and functions are given, we are ready to discuss analysis of those functions. First of all, we have to clarify the meaning of derivative. It may be possible to imitate the derivative in ordinary elementary calculus, but it is not easy. Infinitesimal change of variables which are random should be clarified. Integration should be defined more carefully. Moreover, we shall deal with higher order differential operators. In these cases, one has to be careful about the scale of variables. A simple example is that the white noise $\dot{B}(t)$, the time derivative of a Brownian motion $B(t)$, is of order $\frac{1}{\sqrt{dt}}$. This could illustrate the difficulty of calculus.

The white noise analysis is carried on following the steps proposed above. The system of independent random variables can be taken to be a white noise $\{\dot{B}(t)\}$ which is the time derivative of a Brownian motion $B(t)$. The theory develops successfully through the three steps that are proposed above. This can be seen in the present monograph.

We have so far stated without mathematical rigor, just to explain our basic ideas. From now on we shall develop within the usual mathematical framework.

Basic notions such as probability space $(\Omega, \mathbf{B}, P)$, random variables $X(\omega), Y(\omega)$, ..., finite dimensional probability distributions, etc. are not necessarily defined here. It is the same for the independence so far as finitely many objects are involved. Those notions are familiar to us.

Independence is the key notion of the theory as is understood in the *Reduction* discussed above. We often meet a system of infinitely many (countably many as well as continuously many) independent random variables. We therefore recall, in particular, the concepts in those cases.

A collection $\mathbf{Y} = \{Y_\alpha, \alpha \in A\}$ is an independent system if any finitely many members, say $\{Y_{\alpha_k}, k = 1, 2, \cdots, n\}$ are independent, namely, for any Borel subsets $G_k, 1 \le k \le n$ of R^1, we have the equality

$$P(\bigcap_k Y_{\alpha_k}^{-1}(G_k)) = \prod_k P(Y_{\alpha_k}^{-1}(G_k)).$$

Hence the probability distribution of $\mathbf{Y}$ is the direct product of that of Y_α on R^A. The case where A is continuum, e.g. $A = R^1$, is more interesting and will be discussed carefully. Indeed we shall meet profound properties and see the main interest.

Let $Y_n, n \in N$ be an independent system, where N is the set of natural numbers. We often encounter a convergence problem of a sum of a sequence

$$S = \sum_n Y_n.$$

A sufficient condition for S to be convergence is that both the sum of the means $\sum_n E(|Y_n|)$ and that of variances $\sum_n E(|Y_n - m_n|^2)$ with $m_n = E(Y_n)$ are convergent.

If $S = \sum_n Y_n$ does not converge, but a modified sequence $\sum_n (Y_n - c_n)$ is convergent for some non-random sequence $c_n, n \in N$, then S is said to be *quasi-convergent*.

Related topics in probability theory as well as their variations will be discussed where necessary.

Other basic tools from analysis are

1. Functional analysis. A particular note is that we use the Fréchet derivative for the calculus of functionals.
2. Transformation groups. Infinite dimensional rotation group and symmetric group. They serve to describe various kinds of invariance.

1.2 Discrete parameter white noise

We shall discuss, in this section, the basic time series and its probability distribution. The time series formed by a sequence of independent and identically distributed (i.i.d.) random variables is most important. Let the system be expressed as

$$\cdots, Y(-1), Y(0), Y(1), Y(2), \cdots.$$

If the probability distribution of each $Y(k)$ is standard Gaussian, the sequence $\mathbf{Y} = \{Y(k)\}$ is often called *white noise*, many properties of which are well investigated. However, there are still new properties which should be discovered.

Existence of white noise is proved in many ways. For instance, we show the following examples.

1) Take the Lebesgue measure space $([0,1], \mathcal{B}, dx)$, $\mathcal{B}$ being the Borel field. On this measure space, we have Rademache functions $r_n(x)$, taking values 1 and -1 with probability $\frac{1}{2}$. (For Rademache functions, see the Appendix.) They are independent random variables. Let γ_n be rearranged in a sequence with double indices, say $\mu_{k,l}, 1 \leq k, l \leq \infty$. The functions

$$u_n = \sum_{k=1}^{\infty} \frac{1}{2^{n+1}} \mu_{k,n+1}$$

are independent and subject to the uniform distribution. Take the function $g(x)$ which is the density of the standard Gaussian distribution :

$$g(x) = \frac{1}{\sqrt{2\pi}} e^{-\frac{x^2}{2}}.$$

Set

$$\int_{-\infty}^{x} g(y)dy = G(x).$$

Then, $G^{-1}(u_n) = Y(n)$ is subject to the standard Gaussian distribution. The sequence $\{Y(n)\}$ is exactly what we wish to obtain.

2) Another way of constructing white noise is as follows. Let $Y(k)$ be a random variable defined on a probability space, say

$$\mathbf{\Sigma} = (R, \mathcal{B}, m),$$

with

$$dm(x) = \frac{1}{\sqrt{2\pi}} \exp(-\frac{x^2}{2})dx$$

and $\mathcal{B}$ is the Borel field.

Let $Y = Y(x) = x, x \in R$. Then, $Y(x)$ is a standard Gaussian random variable on Σ. The countably infinite direct product of probability spaces is defined, denoted by $\prod \mathbf{\Sigma}_n$, where $\mathbf{\Sigma}_n$, is a copy of $\mathbf{\Sigma}$.

Define $Y_n(x) = Y(x_n), x = \{x_n\}, x_n \in R$, Then, Y_n is a sequence of i.i.d. random variables, each of which is subject to the standard Gaussian distribution $N(0,1)$. Hence Y_n is a white noise.

Such a construction of white noise leads us to consider a question : how wide is the subset of R^∞, on which the product measure $\prod m_n$ is supported. What we are going to discuss will provide a partial answer to this question.

Our aim is the study of the sequence $\{Y(k), k \in N\}$ which is a system of independent identically distributed, in fact standard Gaussian, random variables and of their functionals. To this end, we must study the probability distribution μ of $\boldsymbol{Y}$.

We discuss characteristic properties of the measure μ.

1) If we take a finite number of $Y(k)$'s, say $Y(k), 1 \le k \le n$, they define the marginal probability distribution μ_k which is n-dimensional standard Gaussian distribution. It is invariant under the n-dimensional rotation group around the origin.

The probability distribution μ of $\boldsymbol{Y} = \{Y(k), k \in N\}$ is not supported by l^2, but is carried by much wider space. This fact can easily be seen, for example, by the strong law of large numbers. It says that

$$\frac{1}{n}\sum_1^n Y(k)^2 \to 1 \tag{1.2.1}$$

almost surely. We see, in particular, μ-almost all points are outside of l^2.

Proposition 1.1 *The series $\sum a_k Y(k)$ converges in $L^2(\Omega, P)$ and also does almost surely if $\sum a_k^2 < \infty$.*

How widely the probability distribution of $\boldsymbol{Y}$ is supported will be illustrated by the Bochner-Minlos theorem which determines, in particular, the probability distribution of $\boldsymbol{Y}$.

Let E be a subspace of the Hilbert space l^2 such that E is equipped with a stronger topology than that of l^2. Namely,

$$E = \{\xi \mid \xi = \{\xi_1, \xi_2, \cdots\}; \sum k^2 \xi_k^2 < \infty\}.$$

For $\xi \in E$ define the norm $\|\xi\|_1$ such as

$$\|\xi\|_1 = \sqrt{\sum k^2 \xi_k^2}.$$

It is easily seen that $\|\cdot\|_1$ is a Hilbertian norm. Thus, E is a topological space dense in l^2.

We can see that the injection T:

$$T : E \to l^2$$

is of Hilbert-Schmidt type.

The dual space of E with respect to the norm in l^2 is denoted by E^*. The norm on E^* is given by

$$\|x\|_{-1} = \sqrt{\sum \frac{1}{k^2} x_k^2}$$

The two spaces E and E^* are linked by the canonical bilinear form :

$$\langle x, \xi \rangle = \sum_i x_i \xi_i, \quad x = (x_i) \in E^*, \xi = (\xi_i) \in E.$$

Note that if x is in l^2, then $\langle x, \xi \rangle$ is just the inner product in l^2.

Then we have

$$E \subset l^2 \subset E^*.$$

The injections $T : E \longmapsto l^2$, $T^* : l^2 \longmapsto E^*$ should be of the Hilbert-Schmidt type and the Hilbert-Schmidt norm is

$$\|T\|_2 = \|T^*\|_2 = \sqrt{\sum_{k=1}^{\infty} \frac{1}{k^2}}.$$

On the other hand, for $\xi \in E$, we may write

$$\langle \boldsymbol{X}, \xi \rangle = \sum X_k \xi_k, \quad \boldsymbol{X} = (X_k),$$

and $\{X_k\}$ is i.i.d. $N(0,1)$.

It can be seen that for all $\xi \in E$, $\langle \boldsymbol{X}, \xi \rangle$ certainly exists and is Gaussian $N(0, \|\xi\|^2)$. (Note that $\|\cdot\|$ is the l^2-norm.)

The characteristic function of finitely many $Y(k)$'s, say $Y(1), \cdots, Y(n)$, is given by

$$E(e^{i\sum_1^n \xi_k Y(k)}) = e^{-\frac{1}{2}\sum \|\xi\|_k^2}$$

where $\xi = (\xi_1, \cdots, \xi_k) \in R^k$ and $\| \|_k$ is the R^k-norm. This is easily defined as a generalization of a characteristic function of one-dimensional random variable.

2) We now come to the case of an infinite dimensional vector.

Proposition 1.2 *The functional $C(\xi)$ of $\xi \in E$:*

$$C(\xi) = E(e^{i\langle X, \xi\rangle}) = \exp\big[-\frac{1}{2}\|\xi\|^2\big],\ \xi \in E \tag{1.2.2}$$

is defined, where $\| \|$ is the l^2-norm. It is a characteristic functional in the sense that

i) $C(\xi)$ is continuous in $\xi \in E$,

ii) $C(\xi)$ is positive definite,

iii) $C(0) = 1$.

Proof. Since $\sum_1^n X_k \xi_k$ converges to $\langle \boldsymbol{X}, \xi\rangle$, almost surely, the bounded convergence theorem proves that

$$\begin{aligned} C(\xi) &= \lim_n E\left(e^{i\sum_1^n X_k \xi_k}\right) \\ &= \lim_n \exp\big[-\frac{1}{2}\sum_k^n \xi_n^2\big] \\ &= \exp\big[-\frac{1}{2}\|\xi\|^2\big]. \end{aligned} \tag{1.2.3}$$

Since $\|\xi\|$ is a continuous l^2-function, the continuity is proved. Positive definiteness comes from the fact that the functional $C(\xi), \xi \in l^2$ is, actually in this case, defined by a probability distribution of $\boldsymbol{Y}$.

The functional $C(\xi)$, given by (1.2.2), is a functional of ξ, continuous in l^2, positive definite and $C(0) = 1$, so that there exists a probability measure μ on E^*, an extension of l^2 by Bochner-Minlos theorem, which we shall prove in Section 1.7 in general setup.

The functional (1.2.3) defines μ which is the standard Gaussian measure on E^* in the following sense.

Take a base $\{e_n\}$ of l^2. Then, for $\xi = \sum_{j=1}^n a_j e_j$, the linear function $\langle x, \xi \rangle$ of $x \in E^*(\mu)$ is a Gaussian random variable with mean 0 and variance $\sum_{j=1}^n a_j^2$. Further, for $\eta = \sum_{j=1}^n b_j e_j$, we see that the covariance of $\langle x, \xi \rangle$ and $\langle x, \eta \rangle$ is $\sum_{j=1}^n a_j b_j$.

Such a simple observation proves that if a and b are orthogonal vector in l^2, then $\langle x, a \rangle$ and $\langle x, b \rangle$ are independent.

1.3 Invariance of the measure μ

What have been discussed lead us to consider the invariant property of the standard Gaussian measure μ, under the transformations acting on E^*, in particular rotation invariance.

Take a finite dimensional subspace $V_n = \{(x_1, x_2, \cdots, x_n); (x_1, x_2, \cdots) \in E\}$. Then we have the marginal distribution μ_n of μ and the characteristic function φ_n of μ_n is of the form

$$\varphi_n(\xi) = e^{-\frac{1}{2}(\xi_1^2 + \cdots + \xi_n^2)},$$

where $\xi = (\xi_1, \cdots, \xi_n) \in F_n$ and where F_n is the n-dimensional subspace of E.

The $\varphi_n(z)$ is a characteristic function on F_n, so that the Bochner theorem on the finite dimensional case, there exists a probability measure μ_n such that

$$C(\xi) = \int_{V_n} e^{\langle x, \xi \rangle} d\mu_n. \tag{1.3.1}$$

Also, by the general theory shown in Section 1.7, μ_n is viewed as the marginal distribution of μ. We therefore prove that any marginal distribution is rotation invariant.

We can see that the marginal distribution μ_n is invariant under rotations of V_n. The rotations of V_n form a group G_n, isomorphic to $SO(n)$.

We now come to the general theory to discuss the invariance of the measure μ, which is the probability distribution of $\boldsymbol{Y}$, the probability of which is μ, the characteristic functional is

$$C(\xi) = \exp[-\frac{1}{2}\|\xi\|^2, \quad \xi \in E.$$

Take a nuclear space E such that E is a dense subspace of l^2. First, we give the definition of rotation group. In fact, the invariance property can be expressed by the rotation group acting on E.

Definition 1.1 Let g be a continuous linear isomorphism of E. If l^2-norm is invariant under g, then g is called rotation of E.

It can easily be seen that the set of all rotations of E forms a group and is denoted by $O(E)$. It is a group topologized by the compact-open topology.

Definition 1.2 The group $O(E)$ consisting of all rotations of E is called the rotation group of E.

If E is not specified in particular, it is called an infinite dimensional rotation group and is denoted by O_∞.

For any $g \in O(E)$ and $\xi \in E$, the map $\xi \mapsto \langle x, g\xi \rangle$ is a continuous linear functional on E. Hence, there exists an element x^* such that $\langle x, g\xi \rangle = \langle x^*, \xi \rangle$. The x^* is uniquely determined by x and g. Hence, we may write $x^* = g^* x$, Thus, we have

$$\langle x, g\xi \rangle = \langle g^* x, \xi \rangle, \tag{1.3.2}$$

where $g^* x \in E^*$. That is, the mapping g^* :

$$x \mapsto g^* x$$

defines a linear isomorphism of E^*.

Set

$$O^*(E^*) = \{g^*, g \in O(E)\}.$$

Theorem 1.1 *We have*

i) The mapping $g^* \mapsto g^{-1}, g \in O(E)$ defines an isomorphism between $O(E)$ and $O^*(E^*)$.

ii) The measure μ is invariant under $g^* \in O^*(E^*)$

Proof. We give the proof as follows.

i) Let $\langle \cdot, \cdot \rangle$ be the canonical bilinear form connecting E and E^*.

Set $\langle g\xi, x \rangle = f(\xi)$. Since $g : \xi \mapsto g\xi$ is continuous in ξ and $\langle \xi, x \rangle$ is a continuous function of ξ, thus $\xi \mapsto \langle g\xi, x \rangle$ is continuous and linear in ξ.

Hence there exists $x_g^* \in E^*$ such that

$$\langle g\xi, x \rangle = \langle \xi, x_g^* \rangle.$$

Since g is an isomorphism of E, the element x_g^* is uniquely determined and linear in x. Hence we may write $x_g^* = g^* x$. Thus we have proved that g^* is a continuous linear automorphism of E^*.

ii) The characteristic functional $C(\xi)$ is invariant under g, so the assertion follows. More precisely, since $d(g^*\mu)(x) = d\mu(g^* x)$, by definition and noting that $(g^*)^{-1} = (g^{-1})^*$, we have

$$\begin{aligned}
\int_{E^*} e^{i\langle x,\xi \rangle} d(g^*\mu)(x) &= \int_{E^*} e^{i\langle x,\xi \rangle} d\mu(g^* x) \\
&= \int_{E^*} e^{i\langle (g*)^{-1} y,\xi \rangle} d\mu(y) \\
&= \int_{E^*} e^{i\langle y, g^{-1}\xi \rangle} d\mu(y) \\
&= e^{-\frac{1}{2}\|g^{-1}\xi\|^2} = C(\xi).
\end{aligned}$$

■

Thus, μ-almost all $y \in E^*$ is viewed as a sample function of $\boldsymbol{Y}$, which is often identify with $y = \{y_k\}$.

The rotation group of n-dimensional space spanned by the system $\boldsymbol{Y} = \{Y(k), 1 \le k \le n\}$ is the same type as in the case of R^n and the probability distribution of $\boldsymbol{Y}$ is invariant under $SO(n)$. The rotation group of the n-dimensional space spanned by $\{Y(k), jn+1 \le k \le (j+1)n\}$ is denoted by G_n^j. It is isomorphic to $SO(n)$ as a topological group.

Set

$$G_n = \bigotimes_n G_n^j \tag{1.3.3}$$

which can give linear transformations of the space generated by the functions of the system $\boldsymbol{Y}$.

The group G expressed in (1.3.3) is isomorphic to a subgroup of $O(E)$, and figuratively speaking, it looks like rotating windmills standing in line. Thus, we may call it **Windmill subgroup**. Details will be discussed later.

More generally, set

$$\boldsymbol{Y_n} = Span\{Y(k), 1 \le k \le n\}. \tag{1.3.4}$$

The rotation group G_n ($\cong SO(n)$) acts on $\boldsymbol{Y_n}$.

Proposition 1.3 *Let G_∞ be the projective limit*

$$G_\infty = \text{proj} \cdot \lim G_n,$$

where the projective limit is taken in the O_∞-topology. Then G_∞ is a subgroup of O_∞.

Note that the projective limit G_∞ of G_n is also isomorphic to a subgroup of $O(E)$.

So far, we are concerned with finite dimensional rotation group. Their actions on E_n and $\boldsymbol{Y_n}$ are bijective, so that the choice of the space is not significant. However, as soon as we come to the entire space E and E^*, their action should strictly be discriminated. We now come to E and E^*.

Proposition 1.4 *For any $g \in G_\infty$, the probability distribution of $g\boldsymbol{Y}$ is in agreement with μ.*

Proof. The characteristic functional of $g\boldsymbol{Y}$ is

$$E[e^{i\langle \xi, gY \rangle}] = E[e^{i\langle g^*\xi, Y \rangle}] = e^{-\frac{1}{2}\|g^*\xi\|^2} = e^{-\frac{1}{2}\|\xi\|^2}.$$

Hence gY has the same probability distributions as μ.

∎

We shall briefly observe the "support" (as it were) of μ in what follows.

i) The series

$$\sum_{k \in Z} k^{-1} Y(k)$$

converges almost everywhere on E^*. If the value of $Y(k)$ is written as y_k, then the series

$$\sum k^{-1} y_k$$

converges for almost all $y = \{y_k\}$.

ii) Take the rotation of the space $\boldsymbol{Y_n}$ defined in (1.3.4). This implies the rotation of $F_n^*(\subset E^*)$ where we have the marginal distribution, denoted by μ_n. That is, we can see the rotation invariance of the measure space (F_n^*, μ_n).

iii) The strong law of large numbers asserts that

$$\lim_{n\to\infty} \frac{1}{N} \sum_1^N Y(k)^2 = 1$$

almost surely. Hence, almost all $y = \{y_n\}$ satisfies

$$\lim_{n\to\infty} \frac{1}{N} \sum_1^N y_n^2 = 1.$$

This means y is outside of l^2. More details are given in Example 1.5.

We may tell the significance of $\boldsymbol{Y}$ in many ways. One of them is seen in innovation theory for time series $X(n)$.

Based on white noise, in other words each $Y(n)$ is taken as a basic variable, we can consider a stochastic process $X(n)$ which is a function of $Y(n)$'s. When we express random developing phenomena by $X(n)$, it is a function of $Y(k)$'s and of the time variable n. Hence, it is important that the so-called **causality** is to be applied. Namely, $X(n)$ is a function of the $Y(k), k \leq n$. Thus, we claim

$$X(n) = \varphi(Y(k), k \leq n; n). \tag{1.3.5}$$

Causality may be well understood in a mathematical communication theory and so it is easy to regard $Y(n)$ and $X(n)$ as input and output, respectively. Equation (1.3.5) means that channel does not forecast the future. In addition, the properties of $X(n)$ can be studied by the calculus on φ and probabilistic properties of white noise $Y(k), k \leq n$. There the analysis is naturally necessary and it should be infinite dimensional analysis.

Here arises a question, "is it possible to obtain φ if $\{X(n)\}$ is given?" . The answer is proposed by P. Lévy[81] rather systematically, i.e. a realization of our idea of "reduction".

1.4 Harmonic analysis arising from $O(E)$ on the space of functionals of $Y = \{Y(n)\}$

A random variable $\boldsymbol{Y}$ is defined on E^* and $x = (\cdots x_{-1}, x_0, x_1, \cdots) \in E^*$ expresses a sample sequence of $\boldsymbol{Y}$. Thus, a function of $\boldsymbol{Y}$ can be expressed as a function of x.

First, consider a complex Hilbert space $L^2(E^*, \mu)$ which is the space of basic white noise functionals with finite variance. We now come to the differential and integral calculus.

The space $\mathcal{H}_1$ generated by all linear functionals of $\{Y(k)\}$ forms a Gaussian system. (General definition of Gaussian system is given in Chapter 2. Here we consider a particular case involving linear functions of $\boldsymbol{Y}$.)

Since we are concerned with Gaussian distribution, we take the Hermite polynomials which form a system of orthogonal basis :

$$H_n(x) = (-1)^n e^{x^2} \frac{d^n}{dx^n} e^{-x^2}, n = 1, 2, \cdots.$$

Let $H_0(x) = 1$.

Example 1.1 If Y is a standard Gaussian random variable, then the collection of

$$\eta_n(Y) = \frac{1}{\sqrt{2^n n!}} H_n(Y/\sqrt{2}), n = 0, 1, 2, ...,$$

forms a complete orthonormal system in the L^2-space spanned by functions of Y with finite variance.

Take a base e_k of E^*. Then, x is expressed as $x = \sum_k x_k e_k$ so the $x_k = \langle x, e_k \rangle$ is the coordinate, and at the same time it is a standard Gaussian variable.

Definition 1.3 A polynomial $\varphi_{\mathbf{k}}(x)$ in $x = (x_1, x_2, \cdots)$ expressed as a finite product :

$$\varphi_{\mathbf{k}}(x) = c_{\mathbf{k}} \prod H_{n_k}(x_k/\sqrt{2}), \tag{1.4.1}$$

on a probability measure space (E^*, μ), is called a *Fourier-Hermite polynomial* with degree $n = \sum n_k$, where $\mathbf{k} = \{k_1, k_2, \cdots\}$ and where $c_{\mathbf{k}}$ is a non-zero constant.

Theorem 1.2 *i) All Fourier-Hermite polynomials with*

$$c_{\mathbf{k}} = (\prod n_k! 2^{n_k})^{-\frac{1}{2}} \tag{1.4.2}$$

is a complete orthonormal system of $L^2(E^*, \mu)$.
ii) Let $\mathcal{H}_n$ be the space spanned by all the Fourier-Hermite polynomials of degree n. Then, they are mutually orthogonal subspaces of $L^2(E^*, \mu)$ and

$$(L^2) \equiv L^2(E^*, \mu) = \bigoplus_0^\infty \mathcal{H}_n \tag{1.4.3}$$

holds.

Proof. i) First we prove that $\{\varphi_n(x)\}$ which are a Fourier-Hermite polynomials are orthogonal.

$$\begin{aligned}
\int \varphi_n(x)\varphi_m(x)d\mu(x) &= c_n c_m \int \prod_j H_{n_j}\left(\frac{\langle x, \xi_j\rangle}{\sqrt{2}}\right) \prod_j H_{m_j}\left(\frac{\langle x, \xi_j\rangle}{\sqrt{2}}\right) d\mu(x) \\
&= c_n c_m \prod_j \int H_{n_j}\left(\frac{\langle x, \xi_j\rangle}{\sqrt{2}}\right) H_{m_j}\left(\frac{\langle x, \xi_j\rangle}{\sqrt{2}}\right) d\mu(x) \\
&= c_n c_m \prod \int H_{n_j}(\frac{x}{\sqrt{2}}) H_{m_j}(\frac{x}{\sqrt{2}}) \frac{1}{\sqrt{2\pi}} e^{-\frac{x^2}{2}} dx \\
&= c_n c_m \delta_{n,m} \prod_j 2^{n_j} n_j!,
\end{aligned}$$

where $m = \{m_j\}$ and $n = \{n_j\}$. See Appendix (A.2.1.3).

Thus, for $n_j \neq m_j$ (for some j), that is for different $\{\varphi_n(x)\}$'s, they are orthogonal. By taking c_j as in (1.4.2), the Fourier-Hermite polynomials are orthonormal.

Since any monomial of the form (1.4.1) as well as any polynomial in $L^2(E^*, \mu)$ can be expressed as a sum of Fourier-Hermite polynomials and the set of polynomials is dense in $L^2(E^*, \mu)$, we can see that $\{\varphi_n(x)\}$ is a base of $L^2(E^*, \mu)$. Thus, the assertion has been proved.

ii) It is easily seen from the definition of $\mathcal{H}_n$ and i). ■

Note for Notation. To clarify the future use of notation, it should be mentioned that $\phi_n(x)$ will denote the n degree orthonormal Hermite polynomials and φ will denote the n degree Hermite polynomials which may not be normalized.

The direct sum decomposition (1.4.3) is called a **Fock space** (Wiener-Itô decomposition of (L^2)).

We now come back to the group $O^*(E^*)$. Since the measure μ is invariant under every g^* in $O^*(E^*)$, we have operators U_g such that

$$(U_g\varphi)(x) = \varphi(g^*x)$$

and

$$\|U_g\varphi\| = \|\varphi\|.$$

We therefore have a unitary representation of $O^*(E^*)$ on $L^2(E^*, \mu)$. Since $O^*(E^*)$ is isomorphic to $O(E)$, we may also say unitary representation of $O(E)$.

Theorem 1.3 *i) $\mathcal{H}_n$ is invariant under U_g for every $g \in O^*(E^*)$.*
ii) The unitary representation U_g of $O(E)$ on the space $\mathcal{H}_n$ is irreducible.

Proof. i) is obvious.
ii) is proved by noting that $O(E)$ has a subgroup G_n which is isomorphic to $SO(n)$.

■

We shall define two transforms that map the space (L^2) to a space of sequences.

Let $\varphi(x)$ be in (L^2, E^*, μ). For $\xi \in E$ define

$$(\mathcal{T}\varphi)(\xi) = \int e^{i\langle x,\xi\rangle}\varphi(x)d\mu(x).$$

This is well defined since $|e^{i\langle x,\xi\rangle}| = 1$. The image of $\mathcal{T}$ is denoted by $\mathcal{F}$.

Proposition 1.5 *The mapping $\mathcal{T}$*

$$\mathcal{T} : (L^2) \longrightarrow \mathcal{F}$$

is linear and bijective, so that $\mathcal{F}$ is topologized so as to be

$$(L^2) \cong \mathcal{F} \ (\simeq \textit{isomorphic})$$

Definition 1.4 The mapping $\mathcal{T}$ is called the $\mathcal{T}$-transform.

In a similar manner we can define $\mathcal{S}$-transform by

$$(S\varphi)(\xi) = C(\xi)\int e^{\langle x,\xi\rangle}\varphi(x)d\mu(x),$$

for $\varphi \in (L^2)$.

The S-transform is well defined since $\|e^{\langle x,\xi\rangle}\| < \infty$ and it is linear.

The two transforms $\mathcal{T}$ and $\mathcal{S}$ are defined in the continuous parameter case and they play important roles in the analysis on the space of continuous parameter white noise functionals.

The following example shows how the S-transform comes from the formula for the Hermite polynomials.

Example 1.2 One-dimensional case.

Let ξ be a real number. The action of the S-transform can be seen from the formula restricted to one-dimensional space. The Hermite polynomial $H_n(\frac{x}{\sqrt{2}})$ is transformed to

$$e^{\frac{-\xi^2}{2}}\int_R e^{\xi x}H_n(\frac{x}{\sqrt{2}})(2\pi)^{-1/2}e^{-x^2/2}dx = (\sqrt{2}\xi)^n. \tag{1.4.4}$$

From this fact follows the proof of the next theorem.

Theorem 1.4 *Let $\varphi_{\mathbf{k}}(x)$ be a Fourier-Hermite polynomial of degree n, expressed in (1.4.1). Its S-transform is expressed in the form*

$$(S\varphi_k)(\xi) = c_k 2^{n/2}\prod_j(\xi_j)^{n_j}, \ \sum_j n_j = n.$$

1.5 Quadratic forms

Quadratic forms in linear algebra are particularly interesting and they have significant properties as is well known. In this section we shall deal with quardatic forms of X_n; we use the notation of the variables X_n instead of Y_n which is used in the previous section.

We can find some characteristic properties of the quadratic forms in infinitely many independent random variables, $X_n, n \in N$, say i.i.d. $N(0,1)$.

Set

$$Q(X_1, X_2, \cdots) = \sum_{i,j} a_{ij} X_i X_j.$$

Since infinite sum is involved, we are concerned with the convergence. A general Q is suggested to be decomposed into two parts : Q_1 and Q_2,

$$Q_1 = \sum_i a_{ii} X_i^2, \quad Q_2 = \sum_{i \neq j} a_{ij} X_i X_j, \tag{1.5.1}$$

where $a_{ij} = a_{ji}$.

As for the convergence in $L^2(\Omega, P)$ of these sums, we can say

i) For Q_1, a_{ii} should be in a trace class.
ii) For Q_2, the condition $\sum_{i \neq j} a_{ij}^2 < \infty$ is requested.

These two cases can be explained as follows.

If we wish to discuss the quadratic forms within the Hilbert space $((L^2) = L^2(\Omega, P))$, we have to recognize the difference between Q_1 and Q_2. For Q_1, the coefficients a_{ii} should be of trace class since $E(X_i^2) = 1$. While $X_i X_j$'s, with $i \neq j$, are orthogonal, the convergence of Q_2 is guaranteed if a_{ij}'s are square summable.

Hence, if we modify Q_1 to be

$$Q_1' = \sum a_{ii}(X_i^2 - 1), \tag{1.5.2}$$

then $\sum a_{ii}^2 < \infty$ is sufficient. Thus, Q_1' belongs to the space (L^2) and the way of convergence is the same for Q_1' and Q_2.

The infinite series Q_1 is said to be *quasi-convergence* (Lévy[84], Chapter 5).

An idea of passage from quadratic form to that of white noise will be discussed in the next chapter.

White noise distributions. Discrete parameter case

I. Quadratic white noise distributions

We are going to introduce a space of generalized quadratic functionals of white noise which is an extension of $\mathcal{H}_2$-space in the Fock space:

$$(L^2) = \bigoplus \mathcal{H}_n,$$

where $\mathcal{H}_n$ is the space spanned by the Hermite polynomials of degree n as is defined in Theorem 1.2. This notation will be used throughout the book.

We shall focus our attention on quadratic functionals which are of particular interest. It is well known that the space $\mathcal{H}_2$ is spanned by the Fourier-Hermite polynomials of degree 2 in the variables X_k's which are independent standard Gaussian random variables. Following the discussions given above those polynomials, having been normalized, are expressed in the form either

$$\psi_k = 2^{-\frac{1}{2}}(Y_k^2 - 1),$$

or

$$\psi_{j,k} = Y_j Y_k, \; j \neq k.$$

Note that ψ_k's and $\psi_{j,k}$'s are members in $\mathcal{H}_2$.

The Hilbert space $\mathcal{H}_2^{(2)}$ of test functionals is defined by

$$\mathcal{H}_2^{(2)} = \{\varphi = \sum a_k \psi_k + \sum b_{j,k}\psi_{j,k}; \sum k^4 a_k^2 + \sum k^2 j^2 b_{j,k}^2 < \infty\}.$$

The dual space (with respect to the L^2-norm) of $\mathcal{H}_2^{(2)}$ is denoted by $\mathcal{H}_2^{(-2)}$, which is called the space of *generalized quadratic functionals* (or quadratic Hida distributions).

We have

$$\mathcal{H}_2^{(2)} \subset \mathcal{H}_2 \subset \mathcal{H}_2^{(-2)}. \tag{1.5.3}$$

The inclusions from left to right, in the above formula, are of Hilbert-Schmidt type.

There are two viewpoints for further analysis of nonlinear functionals of X_n's.

1) For the investigation of general functions of X_n's, we shall take a system of orthogonal functions, since independent systems do not work.
2) Another reason is the following:

   Later, when we wish to come to continuous parameter white noise, we shall observe quadratic forms of continuous parameter as the limit of approximations, which are of the discrete parameter form, for example

$$\sum_{jk} a_{jk} X_j X_k \to \int\int F(u,v) : \dot{B}(u)\dot{B}(v) : dudv,$$

formally. Note that $\dot{B}(t)$'s are continuous analogue of X_n's in a sense. Such an approximation cannot be done easily. In particular, some modification of the term $\sum a_j X_j^2$ is necessary as we shall see the details.

Example 1.3 As a finite dimensional analogue of the standard form of quadratic function, we are interested in

$$\varphi = \sum a_n X_n^2.$$

In particular, take $\varphi_0 = \sum X_n^2$. It is not convergent and even not quasi-convergent. However, $\varphi_1 = \sum (X_n^2 - 1)$ is a generalized quadratic functional in $\mathcal{H}_2^{(-2)}$ as will be shown below.

It is noted that φ_1 is not homogeneous in X_k, but we call it quadratic since it is transformed to $2^{-\frac{1}{2}} \sum_k \xi_k^2$ by applying the S-transform. The same for the case of higher degree.

II. Generalized white noise functionals with discrete parameter

We have discussed generalized quadratic white noise functionals in I, above. We are now in a position to deal with generalized white noise functionals of degree $n(>2)$.

The space $\mathcal{H}_n^+$ is constructed as follows. Take the complete orthonormal system determined in Section 1.4 for $\mathcal{H}_n$ given by Fourier-Hermite polynomials of degree n. Let them be denoted by $\phi_{\boldsymbol{k}}(x)$; $\boldsymbol{k} = (k_1, \cdots, k_m)$ as is mentioned above in "Note for Notation".

A member φ_n in $\mathcal{H}_n$ is expressed in the form

$$\varphi_n = \sum a_{\boldsymbol{k}} \phi_{\boldsymbol{k}},$$

where $\sum |a_{\boldsymbol{k}}|^2 = \|\varphi_n\|^2 < \infty$. Now let the coefficients be more restrictive on a_n, in such a way that

$$\|\varphi_n\|_{n+}^2 \equiv \sum_j \sum_{|\boldsymbol{k}|=n} c_{nj}^2 a_{\boldsymbol{k}}^2 < \infty \tag{1.5.4}$$

with an increasing sequence $c_{nj} > 1$, where $|\boldsymbol{k}| = \sum_j k_j$.

Starting from a vector space L_n spanned by finite sums $\sum a_{\boldsymbol{k}} \varphi_{\boldsymbol{k}}$, we have the completion of L_n with respect to the norm $\| \ \|_{n+}$ to define the space $\mathcal{H}_n^+$.

Take a Gel'fand triple

$$\mathcal{H}_n^+ \subset \mathcal{H}_n \subset \mathcal{H}_n^- \tag{1.5.5}$$

in the usual manner. Then we have a space $(L^2)^-$ of the Hida distributions.

The extension of the Fock space is therefore given by

$$(L^2)^- = \bigoplus c_n \mathcal{H}_n^-, \tag{1.5.6}$$

where $\{c_n\}$ is a sequence of positive decreasing constants. The choice of c_n is arbitrary, but determines the nature of generalized functionals. The topology in $(L^2)^-$ is naturally derived from those of $\mathcal{H}_n^-$ and of c_n. It is noted that we have a freedom to choose a sequence c_n, which can be determined depending on the problem to be discussed.

1.6 Differential operators and related operators

We shall introduce the differential operators with respect to the variables $Y(k)$ coming back to the earlier notation. They are the so-called digital type.

Since $Y(k)$'s are variables of generalized function $\varphi(\mathbf{Y})$, the differentiation with respect to $Y(k)$ is to be defined. The $Y(k)$ is a random variable, so that it is not straightforward to give a definition of differential operator with respect to $Y(k)$. For one thing, one may wonder how to understand a variation of $Y(k)$, denoted by $dY(k)$ which could be non-random or $\epsilon Y(k)$ or something else which is infinitesimal.

Our definition uses the S-transform. There are some reasons to come to this decision. For instance, the S transform of $Y(1)$ or x_1 will be ξ_1. Now the variation $\delta\xi_1$. For another reason, we think of consistency with the case of continuous parameter case that will be given later.

We define the S-transform of $\varphi(Y) = \varphi(Y(k), k \in z)$ by

$$(S\varphi(Y))(\xi) = e^{-\frac{1}{2}\|\xi\|^2} E(e^{\langle Y,\xi\rangle}\varphi(Y)).$$

Before we give the definition of differential operator, we need to prove a proposition.

Proposition 1.6 *The functionals $e^{i\langle x,\xi\rangle}$ and $e^{\langle x,\xi\rangle}$ are both test functionals.*

Proof. For $e^{\langle x,\xi\rangle}$, we modify $e^{\langle x,\xi\rangle}$ to be $e^{t\langle x,\xi\rangle-\frac{1}{2}t^2\|\xi^2\|}$, which is the generating function of the Hermite polynomials with variance $\|\xi\|^2$. Expand it in terms of Hermite polynomials which are in $\mathcal{H}_n^{(n)}$, respectively. ■

The partial differentiation $\frac{\partial}{\partial Y(k)}$ or $\frac{\partial}{\partial x_k}$ with respect to the variable $Y(k)$ or x_k, respectively, is given by

$$S^{-1}\left[\frac{\partial}{\partial \xi_k}(S\varphi)(\xi)\right](Y) = \frac{\partial}{\partial Y(k)}\varphi(Y),$$

$$S^{-1}\left[\frac{\partial}{\partial \xi_k}(S\varphi)(\xi)\right](x) = \frac{\partial}{\partial x_k}\varphi(x),$$

where $\frac{\partial}{\partial \xi_n}$ is the functional derivative.

Note. Some more properties on the S-transform will be given later.

Here the differentiation with respect to x_k, which will be denoted by ∂_k, is defined as follows.

$$\partial_k = \frac{\partial}{\partial x_k} = S^{-1}\frac{\partial}{\partial \xi_k}S.$$

For $\varphi_n(x_j) = H_n(\frac{x_j}{\sqrt{2}})$, using (1.4.4) the partial derivative is

$$\begin{aligned}
\partial_k H_n(\frac{x_j}{\sqrt{2}}) &= S^{-1}\frac{\partial}{\partial \xi_k}\left(S\left(H_n(\frac{x_j}{\sqrt{2}})\right)\right)\\
&= S^{-1}\frac{\partial}{\partial \xi_k}(\sqrt{2}\xi_j)^n\\
&= S^{-1}\delta_{j,k}\sqrt{2}\, n(\sqrt{2}\xi_k)^{n-1}\\
&= \delta_{j,k}\sqrt{2}\, n\, H_{n-1}(\frac{x_j}{\sqrt{2}}). \qquad (1.6.1)
\end{aligned}$$

The result obtained from the above formal calculation is in agreement with the property of Hermite polynomial.

Using the Fock space, we state a theorem.

Theorem 1.5 *The differential operator* $\partial_k = \frac{\partial}{\partial x_k}$ *is*

i) a derivation.

ii) Each $\mathcal{H}_n$ is contained in the domain of ∂_k, so is the Fourier-Hermite polynomial.

iii) ∂_k is the continuous mapping :

$$\mathcal{H}_n^{(n)} \mapsto \mathcal{H}_{n-1}^{(n-1)}.$$

Proof. i) is obvious according to the definition.

ii) The element φ_n of $\mathcal{H}_n$ has an expansion in terms of Hermite polynomials $\varphi_k(x) = H_k(\frac{x}{\sqrt{2}})$ given by

$$\varphi_n = \sum_{\boldsymbol{k}} a_{\boldsymbol{k}} \varphi_{\boldsymbol{k}},$$

where

$$\varphi_{\boldsymbol{k}}(x) = \prod_{j=1}^{m} \varphi_{k_j}(x_j),$$

and where $\boldsymbol{k} = (k_1, k_2, \cdots, k_m)$ such that $k_1 + k_2 + \cdots + k_m = n$. In addition, since φ_n is in (L^2),

$$\sum_{k} \boldsymbol{k}! 2^{|\boldsymbol{k}|} |a_{\boldsymbol{k}}|^2 < \infty,$$

where $\boldsymbol{k}! = \prod_j k_j!$ and $|\boldsymbol{k}| = \sum_{j=1}^{m} k_j$.

According to ∂_k, we have for each term $a_{\boldsymbol{k}}\varphi_{\boldsymbol{k}}$ of the expansion of φ_n,

$$\partial_k \; a_{\boldsymbol{k}} \varphi_{\boldsymbol{k}} = \sum_{j} \delta_{j,k} a_{\boldsymbol{k}} \sqrt{2} \; n \; k_j \; \varphi_{\boldsymbol{k}(j)},$$

where $a_{\boldsymbol{k}}$ is a constant and $\boldsymbol{k}(j)$ means that the factor of φ_j decreases just by one. Obviously, its norm is not greater than $\sqrt{2}\, n$.

iii) For the term $\varphi_{\boldsymbol{k}}$ of the expansion where $k \notin \boldsymbol{k}$, $\partial_k \varphi_{\boldsymbol{k}}$ will vanish, and by ii), we have

$$\|\partial_k \varphi_{\boldsymbol{k}}\| \leq \sqrt{2}\, n \|\varphi_{\boldsymbol{k}}\|.$$

Thus, ∂_k is continuous on the subspace of $\mathcal{H}_{n-1}^{n-1}$ of $(L^2)^+$.

■

Note. Since the partial derivative ∂_k reduces the degree of the Fourier-Hermite polynomial only by one, it is called *annihilation operator*. Note that the term annihilation does not mean that everything has to be vanished.

Since there is an annihilation operator, there must be a creation operator as its adjoint. Actually its existence of the adjoint operator can be shown as follows:

In particular, for $\partial_k \varphi_n(x_k) = \sqrt{2}\, n\varphi_{n-1}(x_k)$, and we have

$$\begin{aligned}\langle \partial_k \varphi_n(x_k), \varphi_{n-1}(x_k)\rangle &= \langle \sqrt{2}\, n\varphi_{n-1}(x_k), \varphi_{n-1}(x_k)\rangle = \langle \varphi_n, \partial_n^* \varphi_{n-1}\rangle \\ &= \sqrt{2}\, n! 2^{n-1} = \langle \varphi_n, \frac{1}{\sqrt{2}}\varphi_n\rangle.\end{aligned}$$

In general, for the Fourier-Hermite polynomials φ and ψ, the bilinear form

$$(\partial_k \varphi, \psi) = (\varphi, \partial_k^* \psi)$$

defines a mapping ∂_k^* which maps the factor φ_{n_k} of ψ to $\frac{1}{\sqrt{2}}\varphi_{n_k+1}$. This relationship can be generalized to that for general Fourier-Hermite polynomials. Note that the case $n_k = 0$ is included.

Theorem 1.6

i) There is an operator ∂_k^* on (L^2) associated with the annihilation operator ∂_j such that

$$\partial_j^* \varphi_{n-1}(x_j) = \frac{1}{\sqrt{2}}\varphi_n(x_j).$$

Consequently each $\mathcal{H}_n$ is contained in the domain of ∂_k^* and

$$\partial_k^* : \mathcal{H}_n \mapsto \mathcal{H}_{n+1}$$

is continuous.

ii) In general, for any φ and ψ in (L^2) the following equality holds.

$$(\partial_k \varphi, \psi) = (\varphi, \partial_k^* \psi).$$

If one side of the above relation is finite, the other side is also finite, and the equality holds.

Proof. i) Let k be fixed and consider the function φ_n of (only) x_k, then

$$\partial_k \varphi_n(x_k) = \sqrt{2}\, n\varphi_{n-1}(x_k).$$

Thus,

$$\langle \partial_k \varphi_n, \varphi_m \rangle = \sqrt{2}\, n \langle \varphi_{n-1}, \varphi_m \rangle = \sqrt{2}\, n \delta_{n-1,m} 2^m m!.$$

Hence, for any $f = \sum a_n \varphi_n(x_k)$,

$$\begin{aligned} \langle \partial_k f, \varphi_m \rangle &= a_{m+1} \sqrt{2}\, (m+1) 2^m m! \\ &= a_{m+1} \frac{1}{\sqrt{2}}\, 2^{m+1} (m+1)! \\ &= \langle f, \frac{1}{\sqrt{2}} \varphi_{m+1} \rangle \\ &= \langle f, \partial_k^* \varphi_m \rangle. \end{aligned}$$

The operator ∂_k^*, defined by

$$\partial_k^* \varphi_m(x_k) = \frac{1}{\sqrt{2}}\, \varphi_{m+1}(x_k),$$

satisfies

$$\partial_k^* \varphi_m(x_j) = \delta_{k,j}\, \frac{1}{\sqrt{2}} \varphi_{m+1}(x_k). \tag{1.6.2}$$

Hence,

$$\partial_k^* : \mathcal{H}_n \mapsto \mathcal{H}_{n+1}$$

and we have

$$\langle \partial_k f, \varphi_m \rangle = \langle f, \partial_k^* \varphi_m \rangle.$$

The proof for a general f follows by taking the coefficient a_n to be the function of variables except x_k, noting that the index k of ∂_k is to be taken arbitrary. Thus, the assertion is proved.

■

Definition 1.5 The operator ∂_k^*, determined by the above theorem, is called the *creation operator*.

Using the calculation of Hermite polynomials we have the following theorem.

Theorem 1.7 *The following commutation relations hold:*

$$[\partial_j^*, \partial_k] = -\delta_{j,k} I,$$

where I is identity mapping and $[\cdot,\cdot]$ denotes the Lie product, i.e. $[A, B] = AB - BA$.

Proof. Using (1.6.1) and (1.6.2), the assertion occurs. ■

Using creation and annihilation operators, define an operator

$$N = \sum_k \partial_k^* \partial_k.$$

Theorem 1.8 *The following facts hold.*

i) The domain of the operator N is the entire space (L^2) and it is a self-adjoint operator.
ii) The eigenvalue of the operator N is non-negative integer and $\mathcal{H}_n$ is the eigenspace belonging to the eigenvalue n.

Proof. A member in $\mathcal{H}_n$ of the Fock space is transformed by the S-transform to a homogeneous polynomial in ξ_k's of degree n. Hence, it is an eigenfunction of $\sum_k \xi_k \frac{\partial}{\partial \xi_k}$. Coming back to the operator acting on $L^2(E^*, \mu)$ and taking S^{-1}, the assertion occurs.

If we change the sign of N to have Δ_∞ which may be thought of as a Laplacian.

Corollary 1.1 *$\mathcal{H}_n$ is the eigenspace of the self-adjoint operator*

$$\Delta_\infty = -\sum_k \partial_k^* \partial_k$$

with eigenvalue $-n$. In particular for $\varphi \in \mathcal{H}_n$

$$\Delta_\infty \varphi = -n\varphi. \tag{1.6.3}$$

In view of this fact, we give the following definition.

Definition 1.6 The operator $N = -\Delta_\infty$ is called the *number operator* because of the relationship (1.6.3), while Δ_∞ is called the *infinite dimensional Laplace-Beltrami operator.*

Proposition 1.7 *The multiplication by x_k is expressed by the sum of ∂_k and ∂_k^*.*

Proof. It can be proved by calculating multiplication by x_k to the Hermite polynomials in x_k. In fact, from the relation

$$2xH_n(x) = H_{n+1}(x) + 2nH_{n-1}(x), \quad x \in R^1,$$

of Hermite polynomials, it is obtained that

$$(\partial_k^* + \partial_k)\varphi(x_k) = x_k\varphi(x_k).$$

Thus, the multiplication p_k by x_k is

$$p_k = \partial_k^* + \partial_k.$$

We can extend this property for general φ.

■

Using the multiplication operators we write

$$-N = \sum_k (\partial_k^2 - p_k\partial_k).$$

This is an infinite dimensional Laplace-Beltrami operator. It can be thought as a kind of infinite dimensional Laplace operator.

There is another Laplacian, called Lévy Laplacian and is denoted by Δ_L. It is defined by

$$\Delta_L = \lim \frac{1}{N} \sum_1^N \partial_k^2$$

still in the discrete parameter case. We do not express it here, but in Chapter 4, we shall discuss the Lévy Laplacian on both the expression and its roles.

Indeed the Lévy Laplacian is quite significant in white noise theory, so we have to consider prudently. Theory of Laplacian can be discussed from many viewpoints, but we have seen some properties based on the computation for functionals of $Y(k)$'s as above. The domain of Δ_L is not discussed here, but we shall determine in the continuous parameter case.

The collection of S-transforms forms a Hilbert space $\mathcal{F}$ (indeed, a reproducing kernel Hilbert space) isomorphic to $L^2(E^*, \mu)$, where we can easily observe operators acting on $\mathcal{F}$ rather than those corresponding operators on $L^2(E^*, \mu)$.

1.7 Probability distributions and Bochner-Minlos theorem

In order to discuss functions $f(X_n, n \in Z)$ (may be better to say functionals of random vector $\boldsymbol{X} = (X_n, n \in Z)$), it is necessary to introduce the probability distribution of $\boldsymbol{X}$.

In order to observe the distribution in question, finite dimensional analogue tells us two reasonable directions (at least).

1. Since X_n's are independent, the probability distribution $\boldsymbol{m}$ of $\boldsymbol{X}$ is a direct product of the one-dimensional standard Gaussian distribution $m = N(0,1)$ introduced on R^1, that is

$$\boldsymbol{m} = \prod_n m_n,$$

with $m_n \equiv m$, for every n.

2. An analogue of the method of characteristic function.

Consider the one-dimensional probability distribution μ. Its characteristic functional $\varphi(z), z \in R^1$ is given by

$$\varphi(z) = \int e^{izx} d\mu(x).$$

If μ is the probability distribution of a random variable X then we have

$$\varphi(z) = E\left(e^{izx}\right).$$

It is well known that there is a one-to-one correspondence between μ and φ, namely the Bochner theorem.

We wish to establish a generalization of the Bochner theorem in the infinite dimensional case, where the random variable is infinite dimensional vector $\boldsymbol{X}$, in somewhat general case. We need to have a clear description of the probability distribution of $\boldsymbol{X}$, to carry on concrete analysis of functionals of $\boldsymbol{X}$, in particular in the case $\boldsymbol{X} = \boldsymbol{Y}$ of the white noise.

An infinite dimensional analogue of a characteristic function is

$$\varphi(\boldsymbol{z}) = E\left(e^{i\langle \boldsymbol{z}, \boldsymbol{X}\rangle}\right),$$

where $\boldsymbol{z} = (z_n; n \in \boldsymbol{Z}, \forall z_n \in R^1)$ and $\boldsymbol{X} = (X_n, n \in \boldsymbol{Z})$.

We now see the convergence of the bilinear form $\langle \boldsymbol{z}, \boldsymbol{X} \rangle = \sum_n z_n X_n$. We assume that X_n's are independent with mean $E(X_n) = 0$ and variance $V(X_n) = 1$, a.e., so that it suffices to prove that $\sum z_n E(X_n)$ and $\sum z_n^2 V(X_n)$ are convergent.

In fact, the former is 0 and the latter is $\sum z_n^2$, so that it suffices to assume $\mathbf{z} \in l^2$ to prove the convergence almost everywhere.

However, continuity of the bilinear form $\langle \boldsymbol{z}, \boldsymbol{X} \rangle$ in $\boldsymbol{z}$ does not hold as we cannot always expect correct evaluation such that

$$|\langle \boldsymbol{z}, \boldsymbol{X} \rangle| \leq \|\boldsymbol{z}\| \sqrt{\sum_j X_j^2},$$

since $\sum_j X_j^2$ may diverge.

Summing up we see that $\varphi(z)$ is not always a continuous function of $z \in l^2$, so that it does not like a characteristic function in the finite dimensional case. Thus, the discussion on the Bochner theorem does not start for $\varphi(\boldsymbol{z})$ in the same manner as in the finite dimensional case.

With this observation, we are requested to establish a generalization of Bochner's theorem to infinite dimensional case with suitable assumptions.

We now recall

Theorem 1.9 *(S. Bochner) Let $\varphi(z), z \in R^n$ be a complex valued function such that*

i) continuous function of $z \in R^n$,
ii) positive definite,
iii) $\varphi(0) = 1$.

Then there exists a probability measure μ on R^n such that

$$\varphi(z) = \int_{R^n} e^{i\langle z,x \rangle} d\mu(x), \tag{1.7.1}$$

where $\langle z, x \rangle$ is the canonical bilinear form that connects R^n and $(R^n)^* \equiv R^n$, i.e. $\langle z, x \rangle = \sum z_i x_i$.

This theorem can be generalized to the infinite dimensional case, with modifications, which are exactly what we are going to clarify.

Let E be a σ-Hilbert nuclear space. The following theorem is known as the Bochner-Minlos theorem (Gel'fand-Vilenkin Generalized Functions, Vol. 4).

Theorem 1.10 *(Bochner-Minlos) Let E be a σ-Hilbert nuclear space. Suppose a functional $C(\xi), \xi \in E$ is*

i) continuous functional of $\xi \in E$,

ii) positive definite, and

iii) $C(0) = 1$.

Then there exists a unique probability measure μ on E^*, the dual space of E, such that

$$C(\xi) = \int_{E^*} \exp[i\langle x, \xi\rangle] d\mu(x). \tag{1.7.2}$$

In this monograph we do not go into such a general theorem, but some restricted case is considered so that existence of probability distributions can be guaranteed visually and their probabilistic properties are clarified rather simply.

Explanation

1) Finite dimensional marginal distributions

We now take n-dimensional subspace of E,

$$F_n = \{(\xi_1, \xi_2, \cdots, \xi_n, 0, 0, \cdots)\} \ (\subset E \subset l^2 \subset E^*)$$

and its dual space

$$F_n^* = \{(x_1, x_2, \cdots, x_n, 0, \cdots)\}$$

are the same. In addition,

$$F \cong R^n \cong (R^n)^* \cong F_n^*,$$

where the notation $\cong$ denotes the isomorphism. Let the projections from E to F_n and E^* to F_n^* be

$$\rho_n : E \mapsto F_n \cong R^n \text{ and}$$
$$\rho_n^* : E^* \mapsto F_n^* \cong R^n,$$

respectively.

Let $C_n(\xi)$ be the restriction of $C(\xi)$ on F_n, which can be considered as the characteristic function $C_n(\xi_1, \xi_2, \cdots, \xi_n)$ defined on R^n. Thus, according to the Bochner theorem for finite dimension, there exists a probability measure m_n on R^n such that

$$C_n(\xi) = \int_{R^n} e^{i\langle x,\xi\rangle_n}\, dm_n(x), \tag{1.7.3}$$

where $\langle\cdot,\cdot\rangle_n$ denotes the inner product on R^n.

Let

$$F_n^a = \{x \in E^*; \langle x, \xi\rangle = 0 \quad \text{for any } \xi \in F_n\}.$$

It can be seen that F_n^a is a vector space of co-dimension n. Form a factor space E^*/F_n^a. Then, we have

$$\langle \tilde{x}, \xi\rangle_n = \langle x, \xi\rangle, \ \xi \in F_n,$$

where $\tilde{x}$ is the class E^*/F^a which contains x. Here $\langle \tilde{x}, \xi\rangle_n$ is a bilinear relation between R^n and $(R^n)^* \cong R^n$. Then, it is

$$F \cong R^n \cong (R^n)^* \cong E^*/F_n^a.$$

By the Bochner theorem, there exists probability measure m_n on F_n^* (or E^*/F_n^a) such that

$$C_n(\xi) = \int_{E^*/F_n^a} e^{i\langle \tilde{x},\xi\rangle_n}\, dm_n(\tilde{x}). \tag{1.7.4}$$

Then, we have a probability measure space $(E^*/F^a, \mathcal{B}_n, m_n)$, where $\mathcal{B}_n$ is a σ-field generated by the Borel subsets of E^*/F_n^a.

2) Consistency of measure spaces and their limit

Let $\rho_{k,n}$, $n > k$, be the projections such that

$$\rho_{k,n} : R^n \mapsto R^k.$$

Then, we have

$$\rho_{k,n}\rho_n = \rho_k, \quad C_k(\rho_{k,n}\xi) = C_n(\xi) \text{ for } \ \xi \in F_n. \tag{1.7.5}$$

From the relation (1.7.5) we have

$$m_k(A) = m_n(\rho_{k,n}^{-1}A), \quad A \in \mathcal{B}_k$$

on the measure space $(R^n, \mathcal{B}_n, m_n)$.

Let $\mathcal{B}$ be the σ-field generated by cylinder sets of E^*.

Next, the limit of $(R^n, \mathcal{B}_n, m_n)$ is to be considered. By using the family of finite dimensional additive sets $\mathcal{A} = \bigcup_n \rho_n^{-1}\mathcal{B}_n$ we can define $(E^*, \mathcal{A}, m)$ such that

$$m(A) = m_n(B) \quad \text{for} \quad A = \rho_n^{-1}(B),\ B \in \mathcal{B}_n.$$

For the extension of this measure space we need deep consideration, which is illustrated in the next step.

3) Extension Theorem

First we will prove the following lemma.

Lemma 1.1 *Let μ_n be a probability measure on R^n and let $\varphi_n(z)$ be the characteristic function of μ_n. Let $U(r)$ be a ball on R^n such that*

$$U(r) = \{z = (z_1, z_2, \cdots, z_n);\ \sum_{i=1}^{n} z_j^2 \le r^2\}$$

and $\beta = (1 - e^{-1/2})^{-1/2}$. For a given $\epsilon > 0$, the inequality

$$|\varphi_n(z) - 1| < \frac{\epsilon}{2\beta^2}$$

holds on $U(r)$.

Let $V(t)$ be an ellipsoid in R^n such that

$$V(t) = \{(x_1, x_2, \cdots, x_n); \sum_{j=1}^{n} a_j^2 x_j^2 < t^2\}, \quad a_j > 0,\ 1 \le j \le n.$$

Then the inequality

$$\mu_n(V(t)^c) < \frac{\epsilon}{2} + \frac{2\beta^2}{r^2 t^2} \sum_{1}^{n} a_j^2 \tag{1.7.6}$$

holds.

Proof. We first prove

$$\begin{aligned} I &= \int_{R^n} \left(1 - e^{-\frac{1}{2t^2}\sum_1^n a_j^2 x_j^2}\right) d\mu_n(x) \\ &\ge \int_{V(t)^c} \left(1 - e^{-\frac{1}{2t^2}\sum_1^n a_j^2 x_j^2}\right) d\mu_n(x) \end{aligned}$$

$$\geq \int_{V(t)^c} \left(1 - e^{-\frac{1}{2}}\right) d\mu_n(x)$$
$$= \frac{1}{\beta^2} \mu_n(V(t)^c).$$

Noting that $e^{-\frac{1}{2t^2}\sum_1^n a_j^2 x_j^2}$ is the characteristic function of a Gaussian measure on R^n, we continue the computation.

$$\int_{R^n} \varphi_n(z) e^{-\frac{t^2}{2}\sum_1^n z_j^2/a_j^2} dz^n = \int_{R^n}\int_{R^n} e^{i\sum_1^n x_j z_j} e^{-\frac{t^2}{2}\sum_1^n z_j^2/a_j^2} dz^n \, d\mu(x)$$
$$= \prod_j a_j \left(\frac{2\pi}{t^2}\right)^{n/2} \int_{R^n} e^{-\frac{1}{2t^2}\sum_1^n a_j^2 x_j^2} d\mu(x).$$

Next,

$$I = \frac{1}{\prod_j a_j}\left(\frac{t^2}{2\pi}\right)^{n/2} \int_{R^n} (1 - \varphi_n(z))\, e^{-\frac{t^2}{2}\sum_1^n z_j^2/a_j^2} dz^n$$
$$|I| \leq \frac{1}{\prod_j a_j}\left(\frac{t^2}{2\pi}\right)^{n/2} \left(\int_{U(r)} |1 - \varphi(z)|\, e^{-\frac{t^2}{2}\sum_1^n z_j^2/a_j^2} dz^n\right.$$
$$\left. + \int_{U(r)^c} |1 - \varphi(z)|\, e^{-\frac{t^2}{2}\sum_1^n z_j^2/a_j^2} dz^n \right)$$
$$< \frac{\epsilon}{2\beta^2} + \frac{1}{\prod_j a_j}\left(\frac{t^2}{2\pi}\right)^{n/2} \frac{2}{r^2} \int_{U(r)^c} \sum_1^n z_j^2 e^{-\frac{t^2}{2}\sum_1^n z_j^2/a_j^2} dz^n$$
$$< \frac{\epsilon}{2\beta^2} + \frac{2}{r^2 t^2}\sum_1^n a_j^2.$$

Accordingly, we have $\mu(V(t)^c) < \dfrac{\epsilon}{2} + \dfrac{2\beta^2}{r^2 t^2}\displaystyle\sum_1^n a_j^2.$

■

Thus, finitely additive measure space $(E^*, \mathcal{A}, m)$ is extended to a completely additive measure space.

Construction of the measure space

We are now in a position to extend the finitely additive measure space $(E^*, \mathcal{A}, m)$. To this end we appeal to the well-known theorem on extendability of a finitely additive.

We now state the lemma in an applicable form in our present situation.

Lemma 1.2 *A finitely additive measure space $(E^*, \mathcal{A}, m)$ extends to a completely additive measure space $(E^*, \mathcal{B}, m)$, $\mathcal{B}$ being the σ-field generated by $\mathcal{A}$, if and only if there exist t and a ball*

$$V(t) = \{x \in E^*, \|x\|_{-1} \le t\}$$

such that $\mu(A) < \epsilon$ for any A and $\epsilon > 0$ with $A \cap V(t) = \varphi$.

Proof. *Necessity*

If an extension μ exists, then there is $V(n)$ such that $\mu(V(n)^c) < \epsilon$. Thus, the assertion holds.

Sufficiency

Suppose $\{A_n\}$ is a partition of E^*. Since m is finitely additive probability measure

$$\sum_{n=1}^{k} m(A_n) \le 1,$$

for any k, so that

$$\sum_{n=1}^{\infty} m(A_n) \le 1.$$

If $\sum_{n=1}^{\infty} m(A_n) < 1$, say

$$\sum_{n=1}^{\infty} m(A_n) < 1 - 3\epsilon < 1,$$

then we know that for any A_n there is an (open) cylinder set A'_n such that $A'_n \supset A_n$ and that $m(A'_n - A_n) < \frac{\epsilon}{2^n}$.

Obviously $\cup A'_n \supset V(n)$. Since a ball is weakly compact, there are finitely many A_n's such that $A' = \cup_1^n A'_n \supset V(n)$ so that

$$m(A') \le \sum_{n=1}^{k} m(A_n) + \epsilon.$$

By assumption

$$m(A'^c) < \epsilon$$

must hold. Hence

$$1 \leq \sum_{n=1}^{k} m(A_n) + \epsilon + \epsilon \leq 1 - 3\epsilon + 2\epsilon = 1 - \epsilon,$$

which is a contradiction.

We now combine what we have discussed. For any $\epsilon > 0$, we take a ball U of l^2 such that for $\xi \in V$,

$$|C(\xi) - 1| < \frac{\epsilon}{2\beta^2}.$$

Take $a_j = 1/j$, $\|x\|_{-1}^2 = \sum_j a_j^2 x_j^2$. Let $V(t)$ be a ball in E^* with radius $t = \frac{2\beta\|T^*\|_2}{r\sqrt{\epsilon}}$. Then, $V(t)$ satisfies the condition of the lemma.

In fact, if the cylinder $A \in \mathcal{A}$ is outside of $V(t)$, and if this cylinder is based on a finite dimensional space F_n, then there is an n-dimensional Borel set $B \in \mathcal{B}_n$ such that

$$A = \rho_n^{-1}(B), \ B \cap \rho_n(V(t)) = \emptyset.$$

Let $V_n(t) = \{x = (x_1, x_2, \cdots, x_n); \sum_{j=1}^n a_j^2 x_j^2 < t^2\}$, then $\rho_n V(t) \subset V_n(t)$.

By Lemma 1.2 and the fact that $|T^*\|_2^2 = \sum a_j^2$,

$$m(A) \leq m_n(\rho_n^{-1} V_n(t)^c) \leq \frac{\epsilon}{2} + \frac{2\beta^2}{r^2 t^2} \sum_1^n a_j^2 < \frac{\epsilon}{2} + \frac{2\beta^2}{r^2} \sum_1^n a_j^2 \frac{\epsilon r^2}{4\beta^2 \|T^*\|_2^2} < \epsilon.$$

By Lemma 1.2, the measure can be extended on E^*. Thus the assertion is proved.

■

Let E be E_1. The functional

$$C(\xi) = e^{-\frac{1}{2}\|\xi\|^2}, \ \xi \in l^2, \tag{1.7.7}$$

is proved to be a characteristic functional. The injection $l^2 \mapsto E^*$ is of Hilbert-Schmidt type. Thus, we have the following theorem.

Theorem 1.11 *The functional $C(\xi)$, $\xi \in l^2$, expressed as above determines the probability measure μ uniquely on $(E^*, \mathcal{B})$ and $C(\xi)$ is the char-*

acteristic functional of the measure μ.

$$C(\xi) = \int_{E^*} \exp[i\langle x, \xi \rangle] \, d\mu(x).$$

Corollary 1.2 *The measure* μ *is the probability distribution of* $\boldsymbol{Y} = \{Y(k)\}$, *the* n*-dimensional marginal distribution is* n*-dimensional standard Gauss distribution.*

Since the measure μ of the space $(E^*, \mathcal{B}, \mu)$ is the distribution of white noise $\boldsymbol{Y}$: the measure space $(E^*, \mathcal{B}, \mu)$ is also called (discrete parameter) **white noise.**

We now discuss the properties of the discrete parameter white noise. For this purpose, we shall observe the probability distribution of the random vector $\boldsymbol{Y}$ defined above, in two ways.

i) The Bochner-Minlos theorem asserts that the probability distribution of $\boldsymbol{Y}$ is introduced on the space E^*, which is larger than l^2, but is much smaller than R^Z which will be defined just below.

ii) Since $Y(k), k \in Z$ is a system of independent random variables, the probability distribution of which is the countably finite direct product of one-dimensional Gauss measure m_k:

$$\mu = \prod_{k \in Z} m_k, \quad m_k = m, \; k \in Z.$$

This μ is defined on R^Z. Let $\mathcal{B}$ be the Borel σ-field on R^Z. Thus the probability space $(R^Z, \mathcal{B}, \mu)$ is constructed.

However, the set supporting μ is not extended over the whole space R^Z. Although the support of μ is not yet defined rigorously, it is to be noted that proper measurable subset A of R^Z is of course very small, intuitively speaking it exists such that $\mu(A) = 1$. The following example tells us this situation.

Example 1.4 The $\{Y(n)^2\}$ is a system of independent random variables with finite mean $E(Y(n)^2) = 1$ and $E(Y(n)^4) = 3$ is finite. By the strong law of large numbers, we have

$$\lim_{N\to\infty} \frac{1}{N} \sum_{k=1}^{N} Y(k)^2 = 1$$

almost surely. Since $x = \{x_n\}$ is a sample point of $\mathbf{Y}$, the set

$$A = \{x; \lim \frac{1}{N} \sum_{1}^{N} x_k^2 = 1\}$$

is measurable and has

$$\mu(A) = 1.$$

We can say that μ is supported by A which is very small compared with R^Z.

This is a different view of the distribution of white noise.

Example 1.5 Application of Kakutani's Dichotomy

For each k, take a slight transition of m. For instance,

$$dm_k(x) = dm(x - a_k),$$

by taking a transition as much as a_k.

If, in particular, we take $a_k = \sqrt{k}$,

$$\sum_k a_k^2 = +\infty$$

holds. This implies that μ and $\mu_{a_k} = \prod m_k$ are mutually singular, intutively speaking, supports of the two measures are disjoint almost surely.

iii) Observation

The volume of n-dimensional sphere with radius r is proportional to r^n. That is, when n gets larger, the volume of the sphere gradually concentrates on the surface. Consider a uniform probability measure on the surface of higher, say n-dimensional sphere. Let the radius of n-dimensional sphere be $\sqrt{n}$. Take a projection of the surface measure down to the straight line passing through the center of the sphere. The corresponding probability distribution, that is the projected measure (marginal distribution), can be computed as follows. See Fig. 1.1.

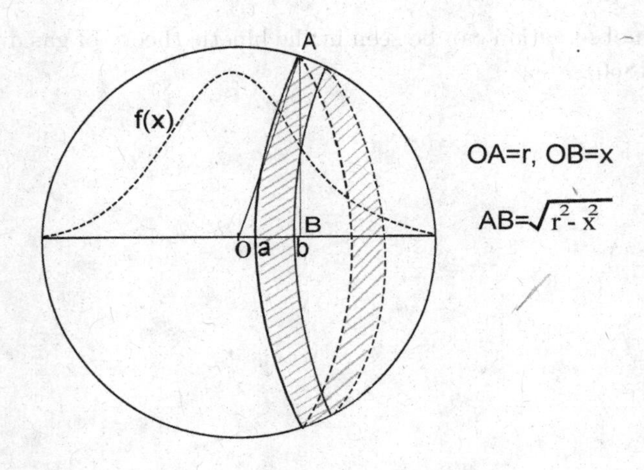

Fig. 1.1

In the above figure $f(x) = \frac{1}{\sqrt{2\pi}} e^{-\frac{1}{2}x^2}$.
Set $r = \sqrt{n}$. Then the shaded area in Fig. 1.1 is

$$\int_a^b (n - x^2)^{\frac{n-3}{2}} \, dx.$$

Take the ratio

$$\frac{\int_a^b (n - x^2)^{\frac{n-3}{2}} \, dx}{\int_{-\sqrt{n}}^{\sqrt{n}} (n - x^2)^{\frac{n-3}{2}} \, dx} = \frac{\int_a^b (1 - x^2/n)^{\frac{n-3}{2}} \, dx}{\int_{-\sqrt{n}}^{\sqrt{n}} (1 - x^2/n)^{\frac{n-3}{2}} \, dx}.$$

If n becomes large, this ratio can be approximated by

$$\frac{\int_a^b e^{-\frac{1}{2}x^2} \, dx}{\int_{-\infty}^{\infty} e^{-\frac{1}{2}x^2} \, dx} = \frac{1}{\sqrt{2\pi}} \int_a^b e^{-\frac{x^2}{2}} \, dx.$$

That is, by taking $n \to \infty$, we obtain the standard Gaussian distribution. This is true for each direction of the sphere in question. Moreover, any projection to higher dimensional space (finite dimensional space, say n-dimensional space) gives the n-dimensional standard Gaussian distribution.

Such an observation can be seen in the kinetic theory of gas due to J.C. Maxwell (1860).

Chapter 2

Continuous Parameter White Noise

Before discussing continuous parameter white noise, we shall note some significant properties of Gaussian system. White noise is a particular and most basic Gaussian, in fact, standard Gaussian system. It is a continuous analogue of i.i.d. idealized random variables, (abbv. i.e.r.v's).

2.1 Gaussian system

In Chapter 1, we have discussed a standard Gaussian system $\{Y(n), n \in Z\}$ parametrized by integers and have developed analysis of their functionals.

Having viewed n as a discrete time parameter, it is natural to introduce a Gaussian system $\{Y(t), t \in R\}$ parametrized by the (continuous) time t. With this system we can describe random evolutional phenomena, in particular, those which are changing as t goes by.

Definition 2.1 $\{X_\lambda, \lambda \in \Lambda\}$ is called a **Gaussian system** if any finite linear combination of X_{λ_k}'s is Gaussian.

Remark 2.1 *1. For convenience to state a fact generally, a constant, say m, is also viewed as a Gaussian random variable with variance 0, i.e. $N(m, 0)$.*

2. Any finitely many random variables from a Gaussian system are subject to a multi-dimensional Gaussian distribution, which may be degenerated.

The probability distribution of Gaussian system $\{X_\alpha, \alpha \in A\}$ is uniquely determined by the expectations $m_\alpha = E(X_\alpha), \alpha \in A$ and the covariance function $\Gamma(\alpha, \beta) = E((X_\alpha - m_\alpha)(X_\beta - m_\beta)), \alpha, \beta \in A$.

We shall state some significant properties of a Gaussian system.

Theorem 2.1 *(P. Lévy-H. Cramér)*
If X and Y are independent and if the sum $X + Y$ is Gaussian, then both X and Y are Gaussian.

This is the well-known theorem, so we omit the proof.

Theorem 2.2 *(P. Lévy)*
If X and Y are such that

$$X = aY + U,$$

where Y and U are independent and a is a constant, and if

$$Y = bX + V$$

where X and V are independent and b is a constant, then there are only three possibilities:

1) X and Y are independent,
2) there is a linear relation between them,
3) $\{X, Y\}$ is a Gaussian system.

Proof is given by using the functional relationships of characteristic functions of random variables involved there.

The following result seems simple, however it has profound meaning concerning a Gaussian system.

Theorem 2.3 *A Gaussian random variable X is atomic in the sense that it does not admit the following decomposition:*

$$X = Y_1 + Y_2, \tag{2.1.1}$$

where Y_1 and Y_2 are mutually independent non-trivial random variable such that $Y_i, i = 1, 2$, are $\mathcal{B}_X$-measurable and $\{X, Y_1, Y_2\}$ is a Gaussian system.

Proof. Suppose we have a decomposition (2.1.1) above. We may assume X, Y_1 and Y_2 have mean 0. Those three random variables form a Gaussian system, so that we have

$$Y_1 = aX + U,$$

where U is independent of X, Then,

$$Y_1 = a(Y_1 + Y_2) + U.$$

Hence we have

$$(1 - a)Y_1 = Y_2 + U.$$

Y_1 is independent of Y_2 and of U, since it is independent of X and so is $\mathcal{B}(X)$ measurable Y_1. Hence the above equation shows that two equal random variables are mutually independent. Hence, we have a trivial random variable, i.e. constant. If $a = 1$, then $U = 0$, i.e. $X = Y_1$. If $a \neq 1$, then Y_1 is constant, that is 0 (hence $X = Y_2$).

Thus, the assertion occurs. ■

Remark 2.2 *Concerning a Poisson random variable X we may also prove that a Poisson random variable is atomic under the same requirements as in the Gaussian case. As a result, we can define the collection $\{\dot{P}(t), t \geq 0\}$, the derivatives of a Poisson process $P(t), t \geq 0$, to be an idealized elemental random variables. But not for a compound Poisson process which is not atomic.*

Now it seems fitting for this chapter to pause for a while in order to explain the main reason of this chapter why we come to a continuous parameter white noise. This fact eventually becomes an interpretation of the aim of this monograph. In fact, we wish to analyze functionals that express random phenomena developing as the time t goes by.

For this purpose we wish to express the given random functions, which are usually complex systems, as functionals of *independent* random variables parametrized by t, let them be denoted by a system $\boldsymbol{Y} = \{Y(t)\}$. The system $\boldsymbol{Y}$ involves continuously many random variables which form a vector space where they are to be linearly independent. In addition, it would be fine if each variable is subject to the same Gaussian distribution.

The first question is to ask if such an ideal system exists. If t is replaced by a discrete parameter like Z, then an ideal system is taken to be $\boldsymbol{Y} = \{Y(k)\}$ as in the previous chapter. The passage from $\boldsymbol{Y}$ to a continuous parameter $\{Y(t)\}$ can never be simple. We have, tacitly, assumed that $\{Y(t)\}$ spans a separable vector space or the probability distribution of the system in question is an abstract Lebesgue space, so that calculus can be proceeded. Such a requirement is quite reasonable from many viewpoints including applications. We therefore have to *exclude* the system, say $\{X(t), t \in R\}$ of independent *ordinary* Gaussian random variables, since such a system cannot satisfy the required assumption.

Thus, an idealized system $\boldsymbol{Y}$ should be such that each $Y(t)$ is elemental (formally speaking, atomic) although it may be an infinitesimal Gaussian random variable and that $Y(t)$'s are equally distributed, maybe in a generalized sense. In addition, they span a separable Hilbert space. We call such a system *idealized elemental Gaussian random variables.*

Our problem is, therefore, to ask if such an idealized system exists. To be quite lucky, we can give an affirmative answer. Formally speaking, the system in question can be given by taking the time derivative of a Brownian motion, denoted by $\dot{B}(t)$ and is called **continuous parameter (Gaussian) white noise**, often called simply as **white noise**.

We now recall the definition of Brownian motion.

Definition 2.2 $\{B(t), t \in R\}$ is a Gaussian system with $E(B(t)) = 0$ and covariance $\Gamma(t,s) = \frac{1}{2}(|t| + |s| - |t-s|)$, then it is called a **Brownian motion**.

By definition it follows that

$$E(B(0)^2) = \Gamma(0,0) = 0$$

which implies $B(0) = 0$, almost surely.

Needless to say, Brownian motion is very important and typical stochastic process not only among Gaussian processes, but also in all stochastic processes. The reason will be clarified in many places in what follows.

We show a method of construction, in fact an approximation of Brownian motion.

Construction of a Brownian motion (white noise)

We have discussed various properties of Brownian motion. We shall now explain the idea of how to construct a Brownian motion $B(t), t \in [0, \infty)$ by the method of interpolation. See Fig. 2.1. We can see the idea of using such interpolation through the property of a Brownian motion stated in Proposition 2.2.

We may assume, theoretically, the existence of a Brownian motion as a Gaussian system.

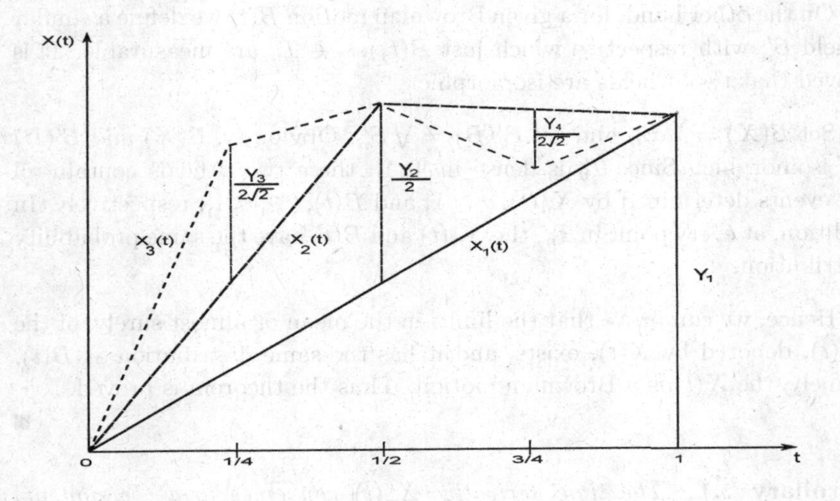

Fig. 2.1

First, let a sequence $\{Y_k, k \geq 1\}$ of independent identically distributed standard Gaussian random variables be provided.

Define a sequence of stochastic processes $X_n(t), t \in [0,1];\ n = 1, 2, \cdots,$ successively as follows.

$$X_1(t) = tY_1. \tag{2.1.2}$$

Let T_n be the set of binary numbers $\frac{k}{2^{n-1}}, k = 0, 1, 2, \cdots, 2^{n-1}$, and set $T_0 = \bigcup_{n \geq 1} T_n$. Assume that $X_j(t) \doteq X_j(t, \omega), j \leq n$, are defined. Then, $X_{n+1}(t)$ is defined in the following manner. At every binary point $t \in T_{n+1} - T_n$ add new random variables Y_k as many as 2^n to $X_n(t)$. On the t-set T_{n+1}^c we have linear interpolation to define $X_{n+1}(t)$.

Then, we claim the following theorem.

Theorem 2.4 *The sequence $X_n(t), n \geq 1$, is consistent in n, and the uniform L^2-limit of the $X_n(t)$ exists. The limit is a version of a Brownian motion $B(t)$.*

Proof. Let $\mathcal{B}_n$ be the σ-field with respect to which all the $X_n(t), t \in [0,1]$. Obviously it is monotone non-decreasing.

On the other hand, for a given Brownian motion $B(t)$ we define a similar σ-field $\mathcal{B}'_n$ with respect to which just $B(t_j), t_j \in T_n$ are measurable. It is proved that two σ-fields are isomorphic.

Set $\mathcal{B}(X) = \bigvee \mathcal{B}_n$ and set $\mathcal{B}'(B) = \bigvee \mathcal{B}'_n$. Obviously, $\mathcal{B}(X)$ and $\mathcal{B}'(B)$ are isomorphic. Since T_0 is dense in $[0,1]$, these two σ-fields contain all the events determined by $X_n(t), n \geq 1$, and $B(t), t \in [0,1]$, respectively. In addition, at every point in T_n, the $X_n(t)$ and $B(t)$ have the same probability distribution.

Hence, we can prove that the limit, in the mean or almost surely, of the $X_n(t)$, denoted by $X(t)$, exists, and it has the same distribution as $B(t)$. Namely, the $X(t)$ is a Brownian motion. Thus the theorem is proved.

■

Corollary 2.1 *The time derivative $X'_n(t)$ converges to a (version of) white noise $\dot{B}(t)$. Note that $X'_n(t)$ is defined except binary points.*

White noise $\dot{B}(t)$ is now understood as a generalized stochastic process in the sense of Gelf'and[18]. Namely, the smeared variable $\langle \dot{B}, \xi \rangle, \xi \in E$, makes sense. We should later give it a rigorous definition to be a generalized variable in $\mathcal{H}_1^{(-1)}$ (in the terminology defined in Chapter 3).

We see further simple properties of Brownian motion.

Proposition 2.1 *If $\{B(t), t \in R^1\}$ is a Brownian motion, then*

1) $E[B(t)B(s)] = \min\{t,s\} \equiv t \wedge s$.
2) $\{\lambda^{-1}B(\lambda^2 t)\}, t \geq 0$ *is a Brownian motion,* $(\lambda \neq 0)$.
3) $B(t)$ *is projective invariance. (The meaning of the projective invariance will be prescribed in the proof.)*

Proof. 1) Take the expectation of

$$B(t)B(s) = \frac{1}{2}\{B(t)^2 + B(s)^2 - (B(t) - B(s))^2\}$$

and use the definition of Brownian motion, the assertion occurs.

2) Let $Y(t) = \lambda^{-1}B(\lambda^2 t)$, then it can be easily seen that $E(Y(0)) = 0$, and

$$\begin{aligned}\Gamma(t,s) = E(Y(t)Y(s)) &= \lambda^{-2}E\left(B(\lambda^2 t)B(\lambda^2 s)\right) \\ &= \lambda^{-2}(\lambda^2 t) \wedge (\lambda^2 s) = \lambda^{-2}\lambda^2(t \wedge s) \\ &= |t| + |s| - |t-s|.\end{aligned}$$

3) The Brownian bridge $X(t)$ over the unit time is defined as

$$X(t) = B(t) - tB(1).$$

We see that its variance and covariance are

$$\begin{aligned}\mathrm{Var}(X(t)) &= t - t^2, \\ \mathrm{Cov}(X(t), X(s)) &= s(1-t), \quad (s < t).\end{aligned}$$

Thus, the normalization of $X(t)$ is

$$Y(t) = \frac{X(t)}{\sqrt{t(1-t)}}$$

and its covariance

$$\mathrm{Cov}(Y(t), Y(s)) = \sqrt{\frac{s}{t} / \frac{1-s}{1-t}}, \quad (s < t)$$

which is anhamonic ratio of $(0, 1; s, t)$.

It is well known that the anharmonic ratio is invariant under projective transformation. Thus, letting p denote the projective transformation, we have

$$(0, 1; s, t) = (p(0), p(1); p(s), p(t)).$$

It shows that the covariance function of $Y(t)$ is invariant under projective invariance. That is, the probability distributions of $\{Y(t)\}$ and $\{Y(p(t))\}$ are the same. Then the assertion holds.

■

Proposition 2.2 *Let $B(t)$ be a Brownian motion and let $0 < a < t < b$. Then*

i) the conditional expectation $B(t; a, b) = E(B(t)|B(a), B(b))$ is interpolation of $B(a)$ or $B(b)$. That is, if $t = \frac{na+mb}{m+n}$, $m, n > 0$, then

$$E(X(t)|B(a), B(b)) = \frac{nX(a) + mX(b)}{m+n}.$$

ii) $B(t) - B(t; a, b)$ *is independent of any* $X(u), u < a$ *and of any* $x(v), v > b$.

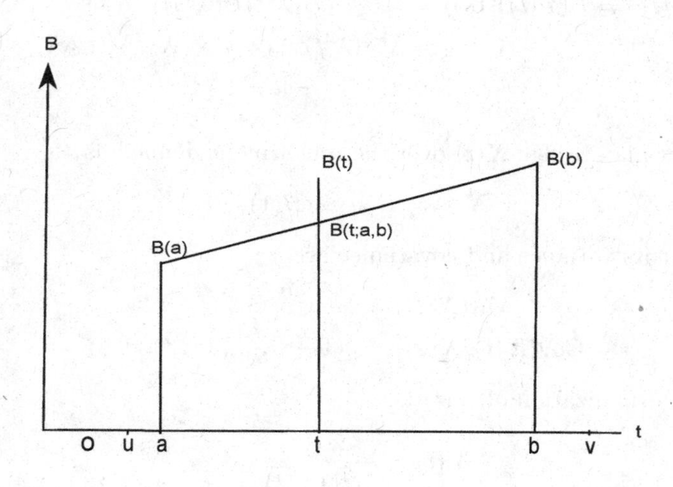

Fig. 2.2

Proof. All the random variables appearing here from a Gaussian system. Hence it suffices to show

$$E(B(t) - B(t; a, b)(\alpha X(a) + \beta X(b))) = 0, \quad \text{for any } \alpha, \beta$$

and to show

$$E(B(t) - B(t; a, b)(\gamma X(u) + \delta X(v))) = 0, \quad \text{for any } \gamma, \delta.$$

Those equalities come from simple computations. ■

It is known that, for almost all ω, the sample function

$$t \mapsto B(t, \omega), \ t \geq 0,$$

is continuous. For the proof, we refer to Hida[27].

We can now come to probability distribution of $\{B(t), t \geq 0\}$. Let $\mathcal{C} = C(0, \infty)$ be the space of continuous functions on $[0, \infty)$, and let

$$A = \{\omega; (\omega(t_1), \cdots, \omega(t_n)) \in \mathcal{B}, \quad 0 \leq t_1 < t_2 < \cdots < t_n\}, \tag{2.1.3}$$

where $\mathcal{B}_n$ is an n-dimensional Borel set. We can introduce a probability distribution μ^W on the measure space $(\mathcal{C}, \mathcal{B})$. Namely, μ^W is the probability measure, on $(\mathcal{C}, \mathcal{B})$, which is uniquely determined by

$$\mu^W(A) = \int \cdots \int_{\mathcal{B}_n} \prod_k \frac{1}{\sqrt{2\pi(t_k - t_{k-1})}} e^{-\frac{y_k^2}{2(t_k - t_{k-1})}} dy_1 \cdots dy_n, \quad (2.1.4)$$

where A is a cylinder set given by (2.1.3).

Definition 2.3 The measure space $(\mathcal{C}, \mathcal{B}, \mu^W(A))$ is called the *Wiener space* and μ^W is called the *Wiener-measure,* respectively.

2.2 Continuous parameter white noise

We have discussed the discrete parameter white noise $\{Y_n(t)\}$ and studied the analysis of its functions. Our next interest which is our main interest is the continuous parameter white noise. As was briefly mentioned before, it is not acceptable to consider a system $\{X_t, t \in R^1\}$ of independently ordinary random variable, say $N(0,1)$ random variables. Apart from this fact, we have a good viewpoint.

Let us think of the relationship between

$$\{Y(n)\} \text{ i.i.d} \longleftrightarrow S(n) = \sum_1^n Y(k).$$

Each side can play almost the same role. The continuous analogue of this fact is

$$dZ(t) \longleftrightarrow Z(t) \quad \text{additive process.}$$

If we assume Gaussian in distributions, we can take a Brownian motion $B(t)$ for $Z(t)$. We wish to replace $dB(t)$ by $\dot{B}(t)$, the time derivative of $B(t)$.

In the previous section, we described white noise $\dot{B}(t)$ as the time derivative of Brownian motion $B(t)$ with respect to time parameter t. However, almost all sample functions $B(t, \omega)$ are not differentiable. Thus the white noise $\{\dot{B}(t), t \in R^1\}$ is defined as a generalized stochastic process. Rigorous definition of a white noise will be given later in Section 2.3.

We are now in a position to discuss the probability distribution of white noise $\{\dot{B}(t), t \in R^1\}$. For a moment white noise is understood as a generalized stochastic process, i.e. the smeared variable $\langle \dot{B}, \xi \rangle, \xi \in E$.

Let E be a nuclear space which is dense in $L^2(R^1)$. It is taken to be the space of test functions. In what follows E is assumed to be one of the following spaces.

i) The Schwartz space S involving C^∞-functions which are rapidly decreasing at $\pm\infty$. It is topologized by the Schwartz topology.
ii) The space $D_0 = \{\xi; \xi(\frac{1}{u})\frac{1}{|u|} \in C^\infty\}$. The space D_0 is isomorphic to $C^\infty(S^1)$.

Each of these spaces is σ-Hilbertian and nuclear. That is, they are topologized by compatible countably many Hilbertian norms $\| \, \|_n$ and satisfy the condition that for any n there exists $n(> m)$ such that the injection

$$T_m^n : E_n \to E_m$$

is nuclear, where E_n is the completion of E with respect to the norm $\| \, \|_n$. In addition, we assume

$$\bigcap_n E_n = E.$$

We shall use these facts.

Take $\xi \in E$. Then we can define the derivative $\dot{B}(t)$ of a Brownian motion in the distribution sense with test function ξ, namely, $\langle \dot{B}, \xi \rangle$. We can show this fact in the following proposition.

We now have a Brownian motion $B(t) = B(t, \omega)$ with parameter $t \in R^1$. Let $\mathcal{B} = \mathcal{B}(B)$ be the sigma-field generated by $\{B(t), t \in R^1\}$.

Proposition 2.3 *For any $\xi \in E, \langle \dot{B}, \xi \rangle$ is defined almost surely and it is $\mathcal{B}$-measurable Gaussian random variable, the probability distribution of which is $N(0, \|\xi\|^2)$. Moreover, $\{B(\xi), \xi \in E\}$ forms a Gaussian system.*

Proof. We observe a sample function which is continuous almost surely. By definition $\langle \dot{B}, \xi \rangle$ is expressed in the form

$$\langle \dot{B}, \xi \rangle = - \int B(t)\xi'(t)dt.$$

Hence,

$$E(\langle \dot{B}, \xi \rangle) = -\int E(B(t))\xi'(t)dt = 0, for\ every\ \xi \in E,$$

$$E(\langle \dot{B}, \xi \rangle \langle \dot{B}, \eta \rangle) = E\left[\int B(t)\xi'(t)dt \int B(s)\eta'(s)ds\right]$$

$$= \int \int (t \wedge s)\xi'(t)\eta'(s)dtds$$

$$= \int \xi(t)\eta(t)dt, \quad \xi, \eta \in E.$$

Thus,

$$Var(\langle \dot{B}, \xi \rangle) = \int \xi(t)^2 dt = \|\xi\|^2.$$

In addition,

$$\sum_j a_j \langle \dot{B}, \xi_j \rangle = \langle \dot{B}, \sum_j a_j \xi_j \rangle$$

is a single bilinear form, so that the sum is a Gaussian random variable.

Thus all the assertions are proved. ■

Note. The random variable $\langle \dot{B}, \xi \rangle$ may be considered as a stochastic integral $\int \xi(t)dB(t) = \int \xi(t)\dot{B}(t)dt$. Such kind of stochastic integral (the Wiener integral) will be discussed in Chapter 5.

With this choice of the test function space E, a white noise $\dot{B}(t)$ through the bilinear form $\langle \dot{B}, \xi \rangle$ is understood in such a way that

i) for almost all $\omega, \dot{B}(\xi)(\omega), \xi \in E$, is a continuous linear functional of ξ,
ii) $\dot{B}(\xi)$ has mean 0 and variance $\|\xi\|^2$, $\|\cdot\|$ is the $L^2(R^1)$-norm,
iii) $\{\dot{B}(\xi), \xi \in E\}$ is a Gaussian system.

In view of Proposition 2.3 we can see that the characteristic functional $C(\xi)$, $\xi \in E$, the expectation of $\exp[i\langle \dot{B}, \xi \rangle]$ of a white noise, is given by

$$C(\xi) = \exp[-\frac{1}{2}||\xi||^2], \ \xi \in E. \tag{2.2.1}$$

So far, we have proceeded in line with the classical theory of stochastic analysis.

Following Hida's idea of *reduction and analysis*, we would like to use $\dot{B}(t)$ itself, without smearing by test function. For one thing, $\dot{B}(t)$ is *elemental*, atomic and mutually independent Gaussian random variable although it is infinitesimal. For another thing, the parameter t stands for the time, so that it can be used to express evolution phenomena that propagate as t changes. Namely, we would like to keep the time variable t not letting it be smeared.

With this requirement in mind, we shall establish a Hilbert space (which will be denoted by $\mathcal{H}_1^{(-1)}$ in Section 3.3), where $\dot{B}(t)$ exists rigorously (with identity) and where $\{\dot{B}(t), t \in R\}$ is a total system.

Before coming to the definition of $\mathcal{H}_1^{(-1)}$, we shall first consider the probability distribution μ of the proposed $\dot{B}(t), t \in R$. The μ-almost all members are considered as sample functions of $\dot{B}(t)$. Hence, we shall have a realization of $\dot{B}(t)$. This realization is given in the following section.

Hence, we should note that the $\langle \dot{B}, \xi \rangle, \xi \in E$, is a generalized stochastic process with characteristic functional (2.2.1) in the sense of Gel'fand[17].

2.3 Characteristic functional and Bochner-Minlos theorem

The rigorous definition of the probability distribution of white noise $\dot{B}(t)$ is given by the Bochner-Minlos theorem.

Let E_0 be $L^2(R^1)$. We start with the Gel'fand triple

$$E \subset E_0 \subset E^*,$$

where the injections are continuous and E^* is the dual space of E.

The probability distribution of a generalized process is a probability measure which is to be defined on E^*.

We denote the canonical bilinear form which links E and E^* by

$$\langle x, \xi \rangle, \quad x \in E, \xi \in E^*.$$

Note that this bilinear form $\langle x, \xi \rangle$ for $x \in L^2(R^1)$ coincides with the inner product on $L^2(R^1)$.

We know that a generalized stochastic process is a system of random variables

$$\{X(\xi,\omega); \xi \in E\}$$

defined on a probability space $(\Omega, \mathcal{B}, P)$ such that $X(\xi,\omega)$ is continuous and linear in ξ. The joint probability distribution of $(X(\xi_1), X(\xi_2), \cdots, X(\xi_n))$ is uniquely determined by the characteristic functional

$$\int e^{i\sum_j z_j X(\xi_j,\omega)} dP(\omega), \quad z_j \in R.$$

This is also the characteristic functional of $X(\sum_j z_j \xi_j)$ since X is linear in ξ.

The joint probability distribution of a generalized stochastic process can be determined by the characteristic functional

$$C_X(\xi) = \int e^{iX(\xi,\omega)} dP(\omega).$$

We can easily see that

1. $C_X(\xi)$ is continuous in $\xi \in E$,
2. $C_X(\xi)$ is positive definite, that is $\sum_{j,k} \alpha_j \overline{\alpha}_k C_X(\xi_j - \xi_k) \geq 0$,
3. $C_X(0) = 1$.

We know that a continuous positive definite function, defined on R^n, determines a probability measure on the same space R^n, however it is not true in the case of Hilbert space. It can be seen in the following example.

Example 2.1 Let $E = L^2(R)$ and

$$C(\xi) = e^{-\frac{1}{2}\|\xi\|^2}, \quad \xi \in E, \ \|\cdot\| : L^2 - \text{norm}. \tag{2.3.1}$$

We can easily verify that $C(\xi)$ satisfies the conditions of characteristic function and thus, the function is a characteristic function. So, we may define the probability measure μ on $E = L^2(R)$. Thus, for any complete orthonormal system $\{\xi_n\}$ in E such that

$$C(z\xi_n) = \int_E e^{iz\langle x,\xi_n\rangle} d\mu(x) = e^{-\frac{z^2}{2}}, \quad z \in R. \tag{2.3.2}$$

But, for any $x \in E$, it must hold that

$$\lim_{n\to\infty} \langle x, \xi_n \rangle = 0,$$

which is in conflict with (2.3.2).

From equation (2.3.2), it can be seen that $\langle x, \xi_n \rangle$ is a standard Gaussian random variable and we have

$$C(\sum_1^n z_k \xi_k) = e^{-\frac{1}{2}\sum_1^n z_k^2}, \quad z_k \in R. \tag{2.3.3}$$

Consequently, $\{\langle x, \xi_n \rangle; n \geq 1\}$ is an independent system of random variables and thus according to the law of large numbers, we have

$$\lim_{N \to \infty} \frac{1}{N} \sum_1^N \langle x, \xi_n \rangle^2 = 1 \quad \text{a.e.}(\mu).$$

Since $\langle x, \xi_n \rangle; n \geq 1$ are thought as the coordinates of x, we can roughly say that the support of μ is much wider than $E = L^2(R)$. This is the reason why the contradiction occurs above.

The probability measure determined by the characteristic functional (2.3.1) is the white noise which is defined later in the following.

Assume that the space E is σ-Hilbert and nuclear. More precisely E is topologized with respect to countably many Hilbertian norms $\| \|_n$ such that E_n, the completion of E by $\| \|_n$ is a Hilbert space and that the injection $E_n \mapsto E_{n-1}$ is of Hilbert-Schmidt type.

$$E = \bigcap_n E_n,$$

where E_0 is the basic Hilbert space $L^2(R^n)$.

We may conclude that

$$\| \cdot \|_0 \leq \| \cdot \|_1 \leq \cdots \leq \| \cdot \|_n \leq \cdots,$$

and they are consistent, i.e. convergence in $\| \cdot \|_n$ implies that in $\| \cdot \|_m, m \leq n$. Then, we have

$$E_0 \supset E_1 \cdots \supset E_n \supset \cdots.$$

Taking $\| \cdot \|_0$ to be the basic norm, the dual space of E_n is denoted by E_n^*, we have

$$E_0^* \subset E_1^* \cdots \subset E_n^* \subset \cdots.$$

Let $\|\cdot\|_{-n}$ be the norm of the Hilbert space E_n^*, the family $\{\|\cdot\|_{-n}, -\infty < n < \infty\}$ is an increasing Hilbertian norms. The dual space E^* can be expressed as

$$E^* = \bigcup_n E_n^*.$$

We now state the Bochner-Minlos theorem for the continuous parameter case in a general setup.

Theorem 2.5 *(Bochner-Minlos) Let $C(\xi), \xi \in E$ be a functional on E which is*

i) continuous in the norm $\| \ \|_p$ for some p,
ii) positive definite, and
iii) $C(0) = 1$.

If for some $n(> p)$ the injection from E_n to E_p is of Hilbert-Schmidt type, then there exists a unique probability measure μ on the measurable space $(E^*, \mathcal{B})$ such that

$$C(\xi) = \int_{E^*} \exp[i\langle x, \xi\rangle] d\mu(x). \tag{2.3.4}$$

This theorem can be proved by using the same technique for the discrete parameter case although the assertion is stated under general setup. For a formal level, we use the fact that a system $\langle x, \xi_n \rangle$ is a coordinate system of x. Namely, we can establish a correspondence between characteristic functionals corresponding to each case. That is, writing $\langle x, \xi_n\rangle = Y_n$,

$$\int e^{i\langle x, \xi\rangle} d\mu(x) \longleftrightarrow E(e^{i\sum a_n Y_n}), \quad a_n = (\xi, \xi_n).$$

Both have the same expression

$$C(\xi) = e^{-\frac{1}{2}\|\xi\|^2}.$$

With such identification, the Bochner-Minlos theorem for continuous parameter case proves the existence and uniqueness of a probability measure μ on E^*.

The Bochner-Minlos theorem uniquely determines the probability distribution μ of white noise.

A cylinder subset of E^* is a set of the form

$$E_n = \{x \in E^*; (\langle x, \xi_1\rangle, \cdots, \langle x, \xi_n\rangle) \in B_n\},$$

where $\xi_1, \cdots, \xi_n, (\in E)$, are linearly independent and B_n is a Borel subset of R^n. The σ-field generated by all the cylinder subsets of E^* will be denoted by $\mathcal{B}$. The measure $\mu(E_n)$ is equal to the Gaussian measure on R^n with mean vector 0 and covariance matrix $((\xi_j, \xi_k))$. Such a measure is a marginal distribution of μ.

Since a white noise $\{\dot{B}(t), t \in R\}$ is a system of generalized random variables, it is difficult to see the geometric structures of the support of μ. But we can prove the following proposition to see some structures.

Proposition 2.4 *The space $L^2(R)$, as a subset of E^*, has μ-measure 0.*

Proof. Take a complete orthonormal system $\{\xi_n\}$ of $L^2(R^1)$ with $\xi_n \in E$, then $\{\langle x, \xi_n\rangle\}$ is a system of independent standard Gaussian distribution on the probability measure space (E^*, μ). Thus, the strong law of large numbers asserts that

$$\lim_{N\to\infty} \frac{1}{N} \sum_1^N \langle x, \xi_n\rangle^2 = 1, \quad \text{a.e. } (\mu).$$

This means $\sum_1^\infty \langle x, \xi_n\rangle^2 = \infty$, i.e. coordinate vector $\{\langle x, \xi_n\rangle\}$ is outside of $L^2(R^1)$.

■

Definition 2.4 The measure space $(E^*, \mathcal{B}, \mu)$ is called white noise.

Note. We have so far called $\{\dot{B}(t)\}$ a white noise in a formal sense. In fact, $\dot{B}(t, \omega)$ is one of the realizations of white noise. Sometimes $\dot{B}(t)$ is also called white noise.

2.4 Passage from discrete to continuous

The advantages to have discrete case are two-fold. One is, as is mentioned in the Preface of Lévy's book (1948), analysis on sure case is applicable and another is used to approximation of continuous case, although we have to be careful on the passage (see 3. below).

Now it is time to compare Theorem 2.5 with Theorem 1.11. The former is concerned with a measure on the function space while the latter discusses the measure on the space of sequences. With a choice of a complete orthonormal system in $L^2(R^1)$ the proof of Theorem 2.5 has been reduced to that of Theorem 1.11.

One should not think that the continuous parameter case can easily be reduced to a discrete parameter case. There are many reasons why we say so.

1. The transition that we did above depends on the choice of a complete orthonormal system ξ_n.
2. The basic spaces, one is a spectrum space and the other is a function space. The latter has very rich and profound structure. As soon as we come to the space of functionals on the measure space, we see that there is given, in the continuous parameter case, a rich class of functionals to be discussed. The same for the operators acting on functionals.
3. If approximation is concerned, we may take $\{\frac{\chi_{\Delta_n}}{\sqrt{\Delta_n}}\}$, $\frac{\Delta_n B}{\Delta_n}$ being a partition of R^1 or interval, which approximates a white noise $\dot{B}(t)$.

Our main aim is first to establish the white noise with continuous parameter by appealing to the Bochner-Minlos theorem (the continuous version). In order to realize the passage, we shall follow the following steps 1), 2) and 3).

The basic function space E is nuclear and is dense in $L^2(R^1)$. Later it will be specified. The dual space of E is denoted by E^* involving generalized functions.

Assume that the characteristic functional $C(\xi), \xi \in E$ is given by

$$C(\xi) = \exp[-\frac{1}{2}\|\xi\|^2].$$

1) Finite dimensional (marginal) distributions

Take a finite dimensional, say n-dimensional subspace F of E.
Define the subspace F^a of E^* by

$$F^a = \{x; \langle x, \xi\rangle = 0 \quad \text{for any } \xi \in F\}.$$

Obviously F^a is a vector space of co-dimension n. Form a factor space E^*/F^a. Then we have

$$E^*/F^a \cong F.$$

As for the bilinear form, we have

$$\langle \tilde{x}, \tilde{\xi} \rangle_F = \langle x, \xi \rangle, \ \tilde{\xi} \in F,$$

where $\tilde{x}$ is a class of E^*/F^a involving the element x. We see that $\langle \tilde{x}, \tilde{\xi} \rangle_F$ is isomorphic to bilinear form between R^n and $(R^n)^* \cong R^n$.

Let $C_F(\xi)$ be the restriction of $C(\xi)$ to F. It is a characteristic function over R^n. Hence by the Bochner Theorem, there is a probability measure m_F on F^* (or E^*/F^a) such that

$$C_F(\xi) = \int_{E^*/F^a} E\left(e^{i\langle \tilde{x}, \tilde{\xi} \rangle_F}\right) dm_F(\tilde{x}).$$

Hence we have a probability measure space $(E^*/F^a, \mathcal{B}_F, m_F)$, when $\mathcal{B}_F$ is the σ-field generated by Borel subsets of E^*/F^a. As was briefly mentioned before, the marginal distribution is always Gaussian.

2) Consistency of measure spaces and their limit

Let ρ_F be the projection such that

$$\rho_F : E^* \mapsto E^*/F^a. \tag{2.4.1}$$

Then, for $G \subset F$, we have

$$\rho_G \, \rho_F = \rho_G. \tag{2.4.2}$$

The family of measure spaces $(E^*/F^a, \mathcal{B}_F, m_F)$ is consistent by the relationship (2.4.2).

Let $\mathcal{B}$ be the σ-field generated by the cylinder subsets of E^*, and set

$$\rho_F^{-1}(\mathcal{B}) = \mathcal{B}_F.$$

We can then define the limit of $(E^*/F^a, \mathcal{B}_F, m_F)$.

It is easy to define a finitely additive measure space $(E^*, \mathcal{A}, m)$, where $\mathcal{A} = \cup_F \mathcal{B}_F$ is a finitely additive field. For the extension of this measure space we need profound consideration, which is illustrated in the next step.

We now specify the choice of a sequence of approximation to the continuous parameter white noise in the following manner.

Let $\{\Delta_k^n, k \in \mathbf{Z}\}$ be a partition of R^1, and take $\{\Delta_k^{n+1}\}$ is finer than $\{\Delta_j^n\}$. Namely, $\{\Delta_k^{n+1}\}$ for any k is a finite sum of Δ_j^n's.

Let (E^*, μ) be a white noise given by Definition 2.4. Take $x \in E^*$.

Set $X_k^n = \langle x, \frac{\chi_{\Delta_k^n}}{\sqrt{\Delta_k^n}} \rangle$. Then $\{X_k^n, k \in \mathbf{Z}\}$ is a system of i.i.d. $N(0,1)$. We wish to say X_k^n tends to $x(t)$ (or to $\dot{B}(t)$) as $\Delta_k^n \to \{t\}$, however this is a formal expression. In reality, we claim the following proposition.

Proposition 2.5 *Suppose* $\sum a_k^n \chi_{\Delta_n}$ *tends to* $f(t)$ *in* $L^2(R^1)$. *Then* $\sum a_k^n \sqrt{\Delta_n} X_k^n$ *tends to* $\langle x, f \rangle$ *in* $L^2(E^*, \mu)$.

Proof. $\{\sum a_k^n \sqrt{\Delta_n} X_k^n, n \geq 1\}$ forms a Cauchy sequence in $L^2(E^*, \mu)$.

Hence the limit exists and is denoted by $\langle x, f \rangle$. It should be noted that we may choose another sequence f_n in $L^2(R^1)$ that converges to f; however the limit is independent of the choice of f_n.

■

3) Observation of the measure space (E^*, μ)

The nuclear space E is now replaced by a function space defined by

$$E = \{\xi \in \mathbf{C}^2(R);\ \|\xi\|_1 \equiv \|D\xi\| < \infty\}, \tag{2.4.3}$$

where $\|\cdot\|$ is the $L^2(R^1)$-norm and where D is a differential operator defined by

$$D = \frac{1}{2}(-\frac{d^2}{du^2} + u^2 + 1).$$

Obviously, $\|\cdot\|_1$ is a Hilbertian norm. Let E_1 be the completion of E by the norm $\|\cdot\|_1$, so that E_1 is a Hilbert space.

Let us pause to review some elementary calculus. The Hermite function $\xi_n(u), n \geq 0$, is defined by

$$\xi_n(u) = \frac{\pi^{-\frac{1}{4}}}{\sqrt{2^n\ n!}} H_n(u) e^{-\frac{u^2}{2}}, \tag{2.4.4}$$

where $H_n(u)$ is the Hermite polynomial of degree n, i.e.

$$H_n(u) = (-1)^n e^{u^2} \frac{d^n}{du^n} e^{-u^2}. \tag{2.4.5}$$

Actual computations tells us that

$$D\xi_n = (n+1)\xi_n, n \geq 0. \tag{2.4.6}$$

We also see that $\{\xi_n\}$ is a base of both $L^2(R)$ and E_1 with a note that $\|\xi_n\| = 1$ and $\|\xi_n\|_1 = n+1$. Thus

$$E_1 = \{\xi; \xi = \sum a_n \xi_n, \ \|\xi\|_1^2 \equiv \sum (n+1)^2 a_n^2 < \infty\}$$

and that $\{\xi_n\}$ is a c.o.n.s of $L^2(R^1)$.

Any ξ in E_1 has an orthogonal expansion: $\xi = \sum a_n \xi_n$ with $\|\xi_n\|_1^2 = \sum (n+1)^2 a_n^2$.

Hence, it is easy to see that the dual space E_1^* of E_1 can be expressed in the form

$$E_1^* = \{x; \ \|x\|_{-1} = \sum \frac{x_n^2}{(n+1)^2} < \infty\}.$$

Note that $x(u)$ may or may not be an ordinary function (i.e. maybe a generalized function). Note that the Hilbertian norm $\|\cdot\|$ is used to define E_1^*.

We now have a Gel'fand triple

$$E_1 \subset L^2(R^1) \subset E_1^*.$$

The injection $E_1 \mapsto L^2(R^1)$ is of Hilbert-Schmidt type, for ξ_n in E_1 has norm $n+1$ and in $L^2(R^1)$ we have unit norm, so that $\frac{1}{(n+1)^2} < \infty$. Similarly the injection $L^2(R^1) \mapsto E_1^*$ is of Hilbert-Schmidt type.

Consider the characteristic functional of white noise $\dot{B}(t)$:

$$C(\xi) = E(e^{i\langle \dot{B}, \xi \rangle}) = e^{-\frac{1}{2}\|\xi\|^2}, \xi \in E_1, \tag{2.4.7}$$

which is continuous in $L^2(R^1)$.

Thus, Gel'fand triple, characteristic functional, Hilbert-Schmidt type injection, etc., is quite isomorphic to the discrete parameter case.

The $C(\xi)$ satisfies necessary conditions for characteristic functional with the choice of Gel'fand triple $E_1 \subset L^2(R^1) \subset E_1^*$.

Hence we have

Theorem 2.6 *The probability distribution μ of white noise $\{\dot{B}(t)\}$ is given on the measure space $(E_1^*, \mathcal{B})$ and*

$$C(\xi) = \int_{E^*} e^{i\langle x,\xi\rangle} d\mu(x)$$

holds, where $\mathcal{B}$ is the σ-field generated by the cylinder subset of E_1^*.

Observations on the support of μ can be done in the same manner to have the same impression.

Warning. By the Proposition 2.4, one may think of μ_σ as a uniform measure on the infinite dimensional sphere (radius $\sigma\sqrt{\infty}$). However, $\{\mu_\sigma, \sigma > 0\}$ cannot be an infinite dimensional analogue of a Lebesgue measure.

We are now ready to define the white noise space.

Definition 2.5 The measure space $(E^*, \mathcal{B}, \mu)$ determined by the characteristic functional (2.4.7) is called a **white noise space**.

Indeed, the white noise space is the probability distribution of $\{B(t); t \in R\}$.

Note. If we take the nuclear space E to be a Schwartz space, then we have Gel'fand triple

$$S \subset L^2(R) \subset S^*.$$

In this case the result of Theorem 2.6, the measure space $(S^*, \mathcal{B}, \mu)$, is also called a white noise space.

Proposition 2.6 *White noise has independent value at every moment in the sense that the random variables $\langle x, \xi_1\rangle$ and $\langle x, \xi_2\rangle$ on $(E^*, \mathcal{B}, \mu)$ are independent whenever $\xi_1(t)\xi_2(t) = 0$. Moreover, $\langle x, \xi_1\rangle$ and $\langle x, \xi_2\rangle$ are independent if and only if $(\xi_1, \xi_2) = 0$.*

Proof of this proposition is straightforward.

We now consider a slightly general case. Set

$$C_\sigma(\xi) = e^{-\frac{\sigma^2}{2}\|\xi\|^2}, \quad \sigma > 0, \xi \in E.$$

Then, there exists measure μ_σ such that

$$C_\sigma(\xi) = \int e^{i\langle x,\xi\rangle} d\mu_\sigma(x).$$

It is easy to see that $\langle x, \xi\rangle$ is a random variable on the probability measure space (E^*, μ_σ). It is Gaussian $N(0, \|\xi\|^2\sigma^2)$. In view of this, the μ_σ is called a Gaussian measure with variance σ^2.

Proposition 2.7 *Two measures μ_{σ_1} and μ_{σ_2} are singular if and only if $\sigma_1 \neq \sigma_2$.*

Proof. Let $\{\xi_n\}$ be a complete orthonormal system in $L^2(R)$ such that $\xi_n \in E$ for every n. Then $\{\langle x, \xi_n\rangle; n = 1, 2, \cdots\}$ forms a sequence of independent, standard Gaussian random variables. Take a set

$$A = \{x; \overline{\lim_{N\to\infty}} \frac{1}{N} \sum_{n=1}^{N} \langle x, \xi_n\rangle^2 = \sigma_1^2\}.$$

The strong law of large numbers gives

$$\mu_{\sigma_1}(A) = 1,$$

while $\mu_{\sigma_2}(A) = 0$. Thus the assertion follows.

■

Thus, we may say that there exist continuously many Gaussian measures on E^*, any two of which are singular.

2.5 Stationary generalized stochastic processes

Since the parameter space is R^1, it allows us to consider the shift operator that defined one-parameter group $\{T_t, t \in (-\infty, \infty)\}$ acting on $R^1 : T_t s = s + t, s \in R^1$. We are therefore suggesting to discuss stationarity of white noise.

Let $X(t), t \in R^1$, be a stochastic process and let $U_h X(t) = X(t+h), t \in R^1$. If $\{U_h X(t)\}$ and $\{X(t)\}$ have the same probability distribution, the $\{X(t)\}$ is called a *stationary* stochastic process.

Consider a Brownian motion $\{B(t) : -\infty < t < \infty\}$. Let $\Delta B(t)$ be the increment $B(t + h) - B(t)$.

By the definition of a Brownian motion, the increment process $\{\Delta B(t), -\infty < t < \infty\}$ is a stationary Gaussian process, and so is white noise $\{\dot{B}(t)\}$. It means that white noise measure μ is invariant under translations by a constant h of the time variable : $t \mapsto t + h$.

The shift of $x \in E^*$ can be defined through the operator S_t of ξ, defined in such a way that

$$(S_t\xi)(u) = \xi(u - t).$$

Each S_t is an automorphism of E and the collection $\{S_t\}$ forms a one-parameter group.

Definition 2.6 If $C(S_t\xi) = C(\xi)$ where C is a characteristic functional, then $(E^*, \mathcal{B}, \mu)$ is called a stationary generalized stochastic process.

Proposition 2.8 *White noise is a stationary generalized stochastic process.*

[**Warning**] Intuitively speaking μ_σ is like a uniformly probability measure on the infinite dimensional sphere $S^\infty(\sigma\sqrt{\infty})$ with radius $\sigma\sqrt{\infty}$. But we cannot imagine $\{\mu_\sigma, \sigma > 0\}$ like a measure equivalent to the Lebesgue measure.

Proposition 2.9 *The time derivative $\dot{X}_n(t)$ of $X_n(t)$, constructed in Section 2.1, converges to a white noise $\dot{X}(t)$ in the sense of weak convergence of stochastic processes, and $\{\dot{X}(t), t \in [0,1]\}$ lives in $\mathcal{H}_1^{(-1)}$, where time parameter is limited to $[0,1]$.*

(Actually $\mathcal{H}_1^{(-1)}$ will be defined in Section 3.3.)

Proof. The characteristic functional $C_n(\xi)$ of $\dot{X}_n(t)$ converges to

$$C(\xi) = \exp(-\frac{1}{2}\int_0^t \xi(t)^2 dt).$$

This can be shown by actual computation.

■

Chapter 3

White Noise Functionals

3.1 In line with standard analysis

Our guiding idea for the white noise analysis is in line with the standard analysis, although the variables are now random and mutually independent. Following this idea, T. Hida has explored a new area in stochastic analysis, where the methods and tools have naturally been proposed from the new viewpoint of the analysis; thus the white noise analysis has been established.

As in the classical calculus, we first provide the variables. Since we are concerned with random phenomena, surely the variables of functions to describe the phenomena should be random. We wish to choose the variables such that they are basic mutually independent and atomic random variables, hopefully Gaussian in distribution. For the analysis of those functions, actually functionals, we would also like to take the Bernoulli's viewpoint, (see the literature[9]) into account, which suggests us to study time developing phenomena.

Now our choice of variables is the white noise[57] $\{\dot{B}(t)\}$. It is a system of *idealized elemental random variables* (abbr. i.e.r.v.), in fact generalized random variables depending on the continuous time parameter t. Rigorous definition together with some interpretation on i.e.r.v. will be given in this chapter. Then, we come to the discussion of functions of white noise $\varphi(\dot{B}(t), t \in R^1)$, variables being parametrized by t, so that actually they are functionals. We start with polynomials in $\dot{B}(t)$'s. So far as linear functionals are concerned, standard technique is applied, although some significant properties in probability theory have been discovered. Apart from this fact a wider class of white noise functionals is introduced so that we can deal with important nonlinear functionals of white noise. It is particularly noted that T. Hida has introduced a technique "renormalization" in probabilistic sense. We follow his technique. To this end, we must keep one thing in mind, namely, $\dot{B}(t)$ is not an ordinary random variable but now it has meaning in the generalized sense.

Moreover, we should seriously think that $\dot{B}(t)$ depends on the time parameter t that runs through an interval which is continuum so that evolutional phenomena can also be expressed explicitly. Also we have not always expect separability of spaces to be introduced because of the uncountable number of variables.

Under the above notes, we will establish the theory of nonlinear functionals of $\dot{B}(t)$'s. With the use of the technique that we have developed, we shall be one further step beyond the classical case.

This step is significant. There are various reasons why we say so. Among them:

i) The member of the base is continuously infinite and not countable. There are important subjects that cannot be well approximated by those in the discrete case.
ii) Brownian motion, hence white noise as well, has countable base and there is so-to-speak separable structure behind.
iii) We use $\dot{B}(t)$'s without smearing, unlike classical theory. As a result, we have generalized functionals of $\dot{B}(t)$'s and with the help of them we can express many significant, even ideal, random phenomena.
iv) Approximation to continuous case by discrete variables requires the clever idea like in physics, however we see difference between our idea and that of physicists.
v) Some visualized interpretations are expected. Our viewpoint can partly be seen through the infinite dimensional rotation group. Naturally follows an infinite dimensional harmonic analysis.

3.2 White noise functionals

Take the measure space $(E^*, \mathcal{B}, \mu)$ established in the previous chapter. Since μ on E^* is the probability distribution of $\{\dot{B}(t); t \in R^1\}$, a member $x \in E^*$ is viewed as a sample function of white noise $\dot{B}(t, \omega) \in E^*$.

Definition 3.1 A complex valued $\mathcal{B}$-measurable function $\varphi(x), x \in E^*$, is called a *white noise functional.*

We can form a complex Hilbert space $(L^2) = L^2(E^*, \mu)$ involving μ-square integrable white noise functionals.

Proposition 3.1 *The algebra $\mathcal{A}$ generated by the system*

$$\{e^{a\langle x,\xi\rangle}; a \in \mathbf{C}, \xi \in E\}$$

is dense in (L^2).

Proof. Let $\varphi \in (L^2)$ be orthogonal to every exponential functional of the form

$$\prod_k e^{it_k\langle x,\xi_k\rangle} \text{ (finite product)},$$

where $\xi_k \in E$ and $t_k \in R$. Then, we have

$$\int_{E^*} \prod_k e^{it_k\langle x,\xi_k\rangle} \overline{\varphi(x)} d\mu(x) = 0.$$

This implies

$$\int_{E^*} \prod_k e^{it_k\langle x,\xi_k\rangle} E(\overline{\varphi(x)}|\mathcal{B}_n) d\mu(x) = 0,$$

where $\mathcal{B}_n$ is the σ-field generated by $\langle x, \xi_k\rangle, 1 \le k \le n$, since the conditional expectation $E(\overline{\varphi}(x)|\mathcal{B}_n)$ is an orthogonal projection of $\varphi(x)$ down to $L^2(E^*, \mathcal{B}_n, \mu)$. Letting t_k run through the real numbers, we have

$$E(\bar{\varphi}(x)|\mathcal{B}_n) = 0, \text{ a.e.}$$

according to the theory of Fourier transform on $L^2(R^n)$. Let $n \to \infty$. Since $\{\xi_n\}$ forms a base of $L^2(R^n)$, we have

$$\varphi(x) = 0, \text{ a.e.}$$

This proves the proposition. ■

Let $P(t_1, \cdots, t_n)$ be a polynomial in $t = (t_1, \cdots, t_n) \in R^n$ with complex coefficients, and let $\xi_1, \cdots, \xi_n$ be linearly independent vectors in E. A functional $\varphi(x)$, expressed in the form

$$\varphi(x) = P(\langle x, \xi_1\rangle, \cdots, \langle x, \xi_n\rangle) \tag{3.2.1}$$

is called a *polynomial* in x. The degree of φ is defined to be that of P.

Recalling that $\langle x, \xi \rangle$ is a Gaussian random variable with mean 0 and variance $\|\xi\|^2 < \infty$, we have

$$\int \langle x, \xi \rangle^{2n} d\mu(x) = (2n-1)!! \|\xi\|^{2n} < \infty, \tag{3.2.2}$$

which proves that any polynomial in x belongs to (L^2).

Proposition 3.2 *The algebra generated by $\{P(\langle x, \xi_1 \rangle, \cdots, \langle x, \xi_n \rangle)\}$ is dense in (L^2).*

A member $\varphi(x)$ in (L^2) is a realization of a functional with a formal expression $\varphi(\dot{B}(t); t \in R^1)$ which is a functional of white noise with finite variance. Thus $\varphi(x)$ in (L^2) is also called a *white noise functional*, since in many cases we tacitly assume μ-square integrability.

Corollary 3.1 *A white noise functional $\varphi(x)$ may be expressed in the form*

$$\varphi(x) = \varphi(\langle x, \xi_1 \rangle, \langle x, \xi_2 \rangle, \cdots)$$

where $\{\xi_n\}$ is a complete orthonormal system in $L^2(R^1)$ such that $\xi_n \in E$ for every n.

Here we note that

$$(\langle x, \xi_1 \rangle, \langle x, \xi_2 \rangle, \cdots)$$

looks like a coordinate representation of x, while $\{\langle x, \xi_n \rangle\}$ is a sequence of independent identically distributed (i.i.d.) standard Gaussian random variables on the probability space $(E^*, \mathcal{B}, \mu)$.

The complex Hilbert space $(L^2) = L^2(E^*, \mu)$ has a representation as the Fock space in the sense that it admits a direct sum decomposition into the spaces of homogeneous chaos $\mathcal{H}_n$, $n = 0, 1, 2, \ldots$, in terms of N. Wiener's,

$$(L^2) = \bigoplus_n \mathcal{H}_n. \tag{3.2.3}$$

More precisely, each subspace $\mathcal{H}_n$ is spanned by the Fourier-Hermite polynomials of the form

$$c \prod_k H_{n_k} \left(\frac{\langle x, \xi_k \rangle}{\sqrt{2}} \right), \quad \Sigma n_k = n,$$

where $\{\xi_k\}$ is a system defined above. Those polynomials are orthogonal to each other in (L^2) and span the entire space. The direct sum decomposition of the form (3.2.3) of the space (L^2) is called the *Fock space*.

A member of a homogeneous chaos $\mathcal{H}_n$ is viewed as a polynomial in white noise. It is also called the *multiple Wiener-Itô integral* of degree n.

Brownian motion can live in the subspace $\mathcal{H}_1$, while $\dot{B}(t)$ is a generalized random variable, so that it is in $\mathcal{H}_1^{(-1)}$ $(\supset \mathcal{H}_1)$ which will be defined later in Section 3.3. It should be noted that a Brownian motion shares the time propagation of random event with the white noise. Roughly speaking, the collection of the events determined by a Brownian motion up to instant t coincides with that of the events determined by the $\{\langle \dot{B}, \xi \rangle, \mathrm{supp}\{\xi\} \subset (-\infty, t]\}$, except the events of probability zero.

In order to come to our main topic we need to introduce a much wider space than (L^2). This will be done in Sections 3.6 and 3.7. Proceeding them, we shall discuss linear functionals of white noise.

3.3 Infinite dimensional spaces spanned by generalized linear functionals of white noise

Linear spaces spanned by $\dot{B}(t)$'s are to be discussed in this section. They might be considered as simple spaces, but not quite. We need to give rigorous definitions and to have them correctly understood.

I. $\mathcal{H}_1^{(-1)}$-space

We first introduce the space $\mathcal{H}_1^{(-1)}$, a space of generalized white noise linear functionals.

The collection of linear functionals of $\dot{B}(t)$'s (later defined as the Wiener integrals in Section 5.2) of the form

$$\dot{B}(f) = \int f(t)\dot{B}(t)dt, \quad f \in L^2(R^1), \tag{3.3.1}$$

spans a Hilbert space $\mathcal{H}_1$ under the $L^2(\Omega, P)$-topology. The above formula proves that the space $\mathcal{H}_1$ is isomorphic to $L^2(R^1)$:

$$\mathcal{H}_1 \cong L^2(R^1). \tag{3.3.2}$$

To prove this fact, we first note that $\dot{B}(\xi) = \int \dot{B}(t)\xi(t)dt$ is defined to be $-\int \dot{B}(t)\xi'(t)$ and it is Gaussian $N(0, \|\xi\|^2)$. The correspondence

$$\dot{B}(t) \longleftrightarrow \xi$$

extends to (3.3.2).

The key point is that the equality

$$E(\dot{B}(f)^2) = \int f(t)^2 dt$$

holds.

Next, let the isometric correspondence

$$f \mapsto \dot{B}(f) \in \mathcal{H}_1, \quad f \in L^2(R^1),$$

extend to that between the Sobolev space $K^{-1}(R^1)$ of order minus one (which is an extension of $L^2(R^1)$) and the space $\mathcal{H}_1^{(-1)}$ which is to be an extension of $\mathcal{H}_1$. Namely, we have a generalization of (3.3.2):

$$\mathcal{H}_1^{(-1)} \cong K^{-1}(R^1). \tag{3.3.3}$$

In reality, $\mathcal{H}_1^{(-1)}$ is defined by (3.3.3).

This isomorphism (3.3.3) determines the $\mathcal{H}_1^{(-1)}$-norm denoted by $\| \ \|_{-1}$. More precisely, if φ $(\in \mathcal{H}_1^{(-1)})$ corresponds to f $(\in K^{-1})$ by (3.3.3) then $\|\varphi\|_{-1} = \|f\|_{K^{-1}(R^1)}$.

We claim that $\dot{B}(t)$ is a member of $\mathcal{H}_1^{(-1)}$ for every t, since $\dot{B}(t)$ can be expressed as $\langle \delta_t, \dot{B}(\cdot)\rangle$ and since $\delta_t \in K^{-1}(R^1)$. The $\mathcal{H}_1^{(-1)}$-norm of $\dot{B}(t)$ is thus obtained by

$$\begin{aligned}
\|\dot{B}(t)\|_{-1}^2 &= \int \frac{|\widehat{\delta_t(\lambda)}|^2}{1+\lambda^2} d\lambda \\
&= \int \frac{|e^{it\lambda}|^2}{1+\lambda^2} d\lambda = \pi.
\end{aligned}$$

The continuity of $\dot{B}(t)$ in t in the space $\mathcal{H}_1^{(-1)}$ is shown by

$$\begin{aligned}
\|\dot{B}(t+h) - \dot{B}(t)\|_{-1} &= \sqrt{\int \frac{|e^{i(t+h)\lambda} - e^{it\lambda}|^2}{1+\lambda^2} d\lambda} \\
&= \sqrt{\int \frac{|e^{ih\lambda} - 1|^2}{1+\lambda^2} d\lambda} \to 0 \text{ as } h \to 0.
\end{aligned}$$

The family $\{\dot{B}(t), t \in R^1\}$ is understood to be a stochastic process in the generalized sense. It is necessary to remind the shift operator S_t and stationarity of a generalized stochastic process is proved with the help of the characteristic functional. Indeed, the $\dot{B}(t)$ is stationary in this sense.

On the space $\mathcal{H}_1^{(-1)}$, the operator U_t is defined based on

$$U_t \dot{B}(s) = \dot{B}(s+t) \tag{3.3.4}$$

and U_t is defined as a linear operator on $\mathcal{H}_1^{(-1)}$. The collection $\{U_t, t \in R^1\}$ forms a one-parameter unitary group which is continuous in t. Hence, the Stone theory is applied to have the spectral decomposition

$$U_t = \int e^{it\lambda} dE(\lambda),$$

where $\{E(\lambda)\}$ is a resolution of the identity. We are now given

$$\dot{B}(t) = U_t \dot{B}(0) = \int e^{it\lambda} dE(\lambda) \dot{B}(0).$$

Set $Z(\lambda) = E(\lambda)\dot{B}(0)$, then

$$\dot{B}(t) = \int e^{it\lambda} dZ(\lambda),$$

and the covariance function is

$$\begin{aligned}
\gamma(h) &= (\dot{B}(t+h), \dot{B}(t))_{\mathcal{H}_1^{(-1)}} \\
&= \int e^{i(t+h)\lambda} e^{-it\lambda} \|dZ(\lambda)\|_{-1}^2 \\
&= \int e^{ih\lambda} E(|Z(\lambda)|^2),
\end{aligned}$$

while

$$\langle \dot{B}(t+h), \dot{B}(t) \rangle_{H_1^{(-1)}} = \int \frac{e^{ih\lambda}}{1+\lambda^2} d\lambda.$$

Hence, we have

$$\|dZ(\lambda)\|_{-1}^2 = \frac{d\lambda}{1+\lambda^2}.$$

Observation

For $\dot{B}(t)$, we have introduced smeared random variable

$$\dot{B}(\xi) = \int \dot{B}(t)\xi(t)dt$$

which has the spectral representation as a stationary random distribution (generalized stochastic process) of the form

$$\dot{B}(\xi) = \int \hat{\xi}(\lambda)dZ(\lambda),$$

and where $\hat{\xi}(\lambda)$ is the Fourier transform of $\xi(t)$.

We can see the following facts:

1) Note that $\dot{B}(t) = \dot{B}(\delta_t)$, the test function being taken to be the delta function, is a well-defined member of $\mathcal{H}_1^{(-1)}$ since δ_t is a member of $K^{-1}(R^1)$. Thus, $\dot{B}(t)$ is given the "exact" (no more formal) **identity**.

2) The space $\mathcal{H}_1^{(-1)}$ is infinite dimensional.
   This is an obvious assertion, since any finite collection of $\dot{B}(t_j), 1 \le j \le n$ (n is arbitrary), with different t_j's, is a system of linearly independent vectors in $\mathcal{H}_1^{(-1)}$. Thus, we may say that there are continuously many linearly independent vectors in the space $\mathcal{H}_1^{(-1)}$.

3) Significant property is that the system $\{\dot{B}(t), t \in R^1\}$ is *total* in $\mathcal{H}_1^{(-1)}$.
   This means that all finite linear combinations of the system span the entire space $\mathcal{H}_1^{(-1)}$ as is claimed in Proposition 3.3 below.
   The above total system plays some role of the system of variables of random functions, although it is not an orthonormal base involving vectors in the ordinary sense.

4) It is noted that the random system $\{\dot{B}(t), t \in R\}$ is *stationary* in time, i.e. its probability distribution μ on E^* (a space of generalized functions on R^1) is invariant under the time shift. In addition, we can prove that the spectrum is *flat*.

As was announced above we prove

Proposition 3.3 *The system $\{\dot{B}(t)\}$ is total in $\mathcal{H}_1^{(-1)}$.*

Proof. We have

$$\mathcal{H}_1^{(-1)} \cong K^{-1}(R^1).$$

This isomorphism asserts

$$\dot{B}(t) \leftrightarrow \delta_t.$$

Each delta function δ_t is a member of $K^{-1}(R^1)$ and there is a dense subset, of $K^{-1}(R^1)$, consisting of the collection $\{\sum_j a_j \delta_{t_j}\}$.

This assertion comes from the fact that $\{\frac{e^{it\lambda}}{\sqrt{1+\lambda^2}}, t \in R^1\}$ is total in $K^{(-1)}(R^1)$.

Thus we have proved that $\{\dot{B}(t)\}$ is total in $\mathcal{H}_1^{(-1)}$.

■

II. $\mathcal{H}_1^{(-n)}$-space

Our next attempt is to have further extension of $\mathcal{H}_1^{(-1)}$, in order to discuss more generalized linear white noise functionals.

Let E_n be the Sobolev space $K^n(R^1)$ of order n over R^1 and let its dual space be denoted by E_{-n}. Namely, there is a Gel'fand triple :

$$E_n \subset L^2(R^1) \subset E_{-n}.$$

We know that $\mathcal{H}_1^{(-1)} \cong K^{(-1)}(R^1)$ which can be generalized in the following line.

Let $K^n(R^1), n > 0$ be the Sobolev space over R^1 of order n. The space $K^{(-n)}(R^1)$ is the dual space of $K^{\frac{n+1}{2}}(R^1)$ with respect to the $L^2(R^1)$-topology.

It is known that the space $\mathcal{H}_1$ of ordinary linear functionals of white noise $\dot{B}(t)$'s is isomorphic to $L^2(R^1)$. The isomorphism can be extended to

$$\mathcal{H}_1^{(-n)} \cong K^{-\frac{n+1}{2}}(R^1), \tag{3.3.5}$$

where $K^{-\frac{n+1}{2}}(R^1)$ is the Sobolev space of order $-(n+1)$ over R^1. This is a generalization of (3.3.3).

Corresponding to the Gel'fand triple, for every $n > 0$,

$$K^{\frac{n+1}{2}}(R^1) \subset L^2(R^1) \subset K^{-\frac{n+1}{2}}(R^1),$$

we have a triple

$$\mathcal{H}_1^{(n)} \subset \mathcal{H}_n \subset \mathcal{H}_1^{(-n)},$$

where each space of which is isomorphic to the corresponding Sobolev space, listed above.

Example 3.1 $B''(t) = \frac{d^2}{dt^2}B(t)$ belongs to $\mathcal{H}_1^{(-2)}$.

3.4 Some of the details of quadratic functionals of white noise

After the spaces of linear functionals we come to quadratic functionals of white noise. As we have seen just before, it is not difficult to deal with linear functionals of $\dot{B}(t)$'s. However, as soon as we come to nonlinear functionals of them we need to introduce a new method of defining them; that is the *renormalization* of familiar functions of the $\dot{B}(t)$'s like polynomials in them. Such a trick is never employed in the classical functionals of smeared variables $\dot{B}(f), f \in L^2(R^1)$. The idea of the renormalization will be first explained for the quadratic forms of $\dot{B}(t)$'s. With this basic observation, we can come to the renormalization of more general non-linear functionals of $\dot{B}(t)$'s in Section 3.10.

We now take quadratic functions (actually nonlinear functionals) of $\dot{B}(t)$'s. General definition of a quadratic functional of white noise is this: an $(L^2)^-$-functional $Q(x) = Q(x(t), t \in R^1)$ is said to be *quadratic* if its S-transform $U(\xi)$ satisfies the following property. For any $\xi, \eta \in E$ and for any $\alpha, \beta \in R^1$, the function $U(\alpha\xi + \beta\eta)$ is a homogeneous polynomial of degree 2 in α and β, then $U(\xi)$ and hence $Q(x)$ is called a quadratic functional or quadratic form.

In general, $U(\alpha\xi + \beta\eta)$ is a homogeneous polynomial of degree n in α and β, then U and Q are called entire homogeneous of degree n.

We claim that among others the subspace $\mathcal{H}_2^{(-2)}$ consisting of *quadratic* generalized white noise functionals is particularly important by many reasons so as to proceed further discussions.

Before we discuss the details of the analysis of quadratic forms of the $\dot{B}(t)$, we shall emphasize the significance of "quadratic". For the discrete parameter case we have made a short note in Section 1.5. The decomposition of a general quadratic form $Q(x)$ as well as its significance can also be mentioned in the continuous parameter case as is seen below.

For the i.i.d. sequence $\{X_n\}$, each X_n being $N(0,1)$ random variable, we may repeat as follows.

We have

$$Q(X) = \sum_{1\le i,j\le n} a_{ij}X_iX_j = Q_1(X) + Q_2(X), \tag{3.4.1}$$

where

$$Q_1(X) = \sum_{i=1}^{n} a_{ii}X_i^2, \tag{3.4.2}$$

$$Q_2(X) = \sum_{1\le i,j\le n, i\ne j} a_{ij}X_iX_j, \ a_{ij} = a_{ji}. \tag{3.4.3}$$

Also we remind the condition for convergence, in particular quasi convergence as $n \to \infty$. Thus we require convergence for $\sum a_{ii}$ and square summability of coefficients a_{ij}. Also, quasi-convergence of $Q_2(x)$ as $n \to \infty$. Such an observation gives a suggestion to have renormalized quadratic forms (see Section 3.8).

We are now ready to discuss *the passage from discrete to continuous* (in Hida's words the passage from digital to analogue).

For our purpose of the passage X_j^n is taken to be the variation of a Brownian motion. (See the Lévy's construction of a Brownian motion by interpolation.) Namely, we set $X_j^n = \frac{\Delta_j^n B}{\sqrt{\Delta^n}}$, where $\{\Delta_j^n\}$ is the partition of $[0,1]$ with $\Delta_j^n = [\frac{j}{2^n}, \frac{j+1}{2^n}]$, hence, with $|\Delta_j^n| = |\Delta^n| = \frac{1}{2^n}$.

Then,

$$Q_2(X) = \sum_{i\ne j} a_{ij}^n \frac{\Delta_i^n B}{\sqrt{|\Delta^n|}} \frac{\Delta_j^n B}{\sqrt{|\Delta^n|}}$$

converges with the assumption

$$\sum (a_{ij}^n)^2 < \infty, \tag{3.4.4}$$

as $n \to \infty$.

The above inequality (3.4.4) guarantees the convergence of

$$\sum_{i\neq j} \frac{(a_{ij}^n)}{|\Delta^n|} \frac{\Delta_i^n B}{|\Delta^n|} \frac{\Delta_j^n B}{|\Delta^n|} |\Delta^n||\Delta^n|,$$

as letting $|\Delta^n| \to 0$ as $n \to \infty$. The sum tends to

$$Q_2(\dot{B}) = \int_0^1 \int_0^1 F(u,v)\dot{B}(u)\dot{B}(v)dudv \tag{3.4.5}$$

in $\mathcal{H}_2$, where $\frac{a_{ij}^n}{\Delta^n}$ approximates $F(u,v) \in \widehat{L^2(R^2)}$ so that

$$\sum_{i\neq j} \frac{(a_{ii}^n)}{\Delta_n^2} \Delta_n \Delta_n \to \int_0^1 \int_0^1 F(u,v)dudv.$$

As for the limit of $Q_1(x)$, we need an additional treatment. First we note that as $\Delta \to \{t\}$ the quantity $(\frac{\Delta B}{\Delta})^2$ may be considered to tend to $\dot{B}(t)^2$, however it is not a generalized white noise functional, which we will introduce in Section 3.8. If we have the difference $(\frac{\Delta B}{\Delta})^2 - \frac{1}{\Delta}$, then it converges to a generalized quadratic functional which is denoted by $: \dot{B}(t)^2 :$. Rigorous discussion will be given in Section 3.8, and for a moment we understand this fact. We must modify $Q_1(x)$

$$Q_1(x) \longmapsto Q_1'(x) = \sum_i a_{ii}(\frac{(\Delta_i^n B)^2}{|\Delta^n|} - \frac{1}{|\Delta^n|}$$

which tends to

$$Q_1(\dot{B}) = \int_0^1 f(u) : \dot{B}(u)^2 : du, \tag{3.4.6}$$

with $f \in L^2(R^1)$.

Such a modification is made following the general rule of renormalization which will be discussed in Section 3.10. The result (3.4.6) is a Hida distribution while (3.4.5) defines an ordinary $\mathcal{H}_2$-functional.

So far we observed some particular generalized functionals of $\dot{B}(t)$ of degree 2, where we have seen the idea of the passage from discrete to continuous. The limit of $Q(X)$, is in fact, *normal functional* in terms of P. Lévy,

$$\int f(u) : \dot{B}(u)^2 : du + \int\int F(u,v) : \dot{B}(u)\dot{B}(v) : dudv. \tag{3.4.7}$$

We just mentioned above that equation (3.4.7) is the normal functionals in $\dot{B}(t)$. The second integral uses the notation ::; this is unnecessary in (3.4.5). Thus, we should now give the general expression of *Lévy's normal functional* in terms of S-transform which is of the form

$$\int f(u)\xi(u)^2 du + \iint F(u,v)\xi(u)\xi(v)dudv, \tag{3.4.8}$$

where $f \in L^1(R), F \in L^2(R^2)$ and $\xi \in E$.

In other words, we can say that a functional is called *normal functional* in Lévy sense if Fréchet's derivatives up to the second order exist and the second Fréchet's derivative has singularities, if any, only on the diagonal.

We now pause to explain why normal functionals are important. Most significant reason is in the analytic property. For this purpose we remind the definition of Fréchet derivative. (We do not take the Gâteaux derivative.)

Definition 3.2 Let A be a bounded linear operator on a Banach space E with norm $\|\cdot\|$. If for a function f on E satisfies

$$\lim_{y\to 0} \frac{\|f(x+y) - f(x) - A(y)\|}{\|y\|} = 0,$$

where $A(y)$ is linear, then f is called *Fréchet differentiable* and A is called *Fréchedt derivative.*

If, in particular, E is a function space, then A is assumed to be an integral operator with kernel function function $A(f,t)$

Note. A Gâteaux differentiable f is Fréchet differentiable if and only if

$$\lim \frac{1}{t}(f(x+ty) - f(x)) = df(x)$$

exists and

$$\sup_{y\neq 0} \frac{\|df(x,y)\|}{\|y\|} < \infty.$$

The second derivative is just to repeat again.

Returning to the normal functional, we may say that the first term of (3.4.7) has the Fréchet derivative and the kernel function has singularity only on the diagonal. We may therefore call it the *singular* part of the normal functional. While the second term is second Fréchet differentiable with kernel function $F(u, v)$. In view of this it is called the *regular* part of the normal functional. The two terms can be discriminated by the Lévy Laplacian (for definition see Section 4.3). The regular part of any normal functional is always harmonic. The singular part belongs to the domain of the Lévy Laplacian.

There are some other reasons why we emphasize the significance of quadratic functionals. As we have seen in the passage from discrete to continuous, the normal functionals appear as the limit.

Some other characteristics of the quadratic functional cân be seen through the representation by using the $\mathcal{T}$ or $\mathcal{S}$ transform. For an ordinary functionals, we are given a representation in terms of symmetric $L^2(R^2)$ function. We can then appeal to Mercer's theorem to develop the original quadratic random functional decomposed into the sum of countably many independent random variables with χ^2-distribution. Good example is the Lévy's stochastic area discussed in the next section.

Quadratic forms can be a good example to see a plausibility of renormalization. Reductionism, which is Hida's favorite idea, suggests us to start the analysis with the algebra of polynomials in $\dot{B}(t)$'s. Except linear functionals, those polynomials need renormalization in order to be generalized functionals of $\dot{B}(t)$'s. Quadratic form that is the limit of $Q_1(x)$ shows why and how the renormalization is necessary to be a generalized quadratic functional. We emphasize that such a generalized functional plays significant role also in the applications in physics.

Finally, when Laplacian is discussed, normal quadratic forms illustrate how it works from the view point of "harmonic analysis", where infinite dimensional rotations are acting.

The shift S_t

In the previous section we have briefly mentioned the shift acting on $\dot{B}(t)$'s. The unitary operator U_t on $\mathcal{H}_1^{(-1)}$ has also been discussed. We now come to the shift in a general setup.

Define a transformation S_t :

$$(S_t\xi)(u) = \xi(u - t).$$

It is easy to see that S_t is a member of $O(E)$, and we have

$$S_t S_s = S_{t+s}, S_0 = I.$$

Also we prove

$$S_h\xi \to \xi \text{ in } E \text{ as } h \to 0.$$

Thus, $\{S_t, t \in R\}$ is a one-parameter subgroup of $O(E)$ and continuous in t.

We now think of the characterization of normal functionals. Among others the following two properties are to be mentioned.

1) The spectrum of the shift.
   On the space spanned by Q_1-type S_t has continuous spectrum of the Lebesgue type and unit multiplicity.
   On the other hand, on the space spanned by the Q_2-type S_t has Lebesgue spectrum and the multiplicity is countably infinite.

2) The harmonic property can be discussed after the Lévy Laplacian Δ_L is defined (see Section 4.3). Both Q_1 and Q_2 have enough analytic properties, in particular they are in the domain of Δ_L. We see, as mentioned before, that Q_2 is always harmonic, while Q_1 may have non-zero trace.

As for the renormalization of quadratic functionals, we shall first tell the result. There is an isomorphism

$$\mathcal{H}_2^{(-2)} \cong \widehat{K^{-3/2}(R^2)}, \tag{3.4.9}$$

as an extension of the known isomorphism

$$\mathcal{H}_2 \cong \widehat{L^2(R^2)}. \tag{3.4.10}$$

According to the isomorphism (3.4.9), for $\varphi \in \mathcal{H}_2^{(-2)}$ there is a function $F(u,v)$ in the space $\widehat{K^{-3/2}(R^2)}$ to have the representation

$$\varphi(\dot{B}) = \int F(u,v) : \dot{B}(u)\dot{B}(v) : dudv, \tag{3.4.11}$$

where the notation $: \cdot :$ may be considered as the Wick product, i.e. renormalized product.

3.5 The $\mathcal{T}$-transform and the S-transform

In this section we use the notation x instead of $\dot{B}(t)$.

The $\mathcal{T}$-transform and the S-transform play an essential role in Hida distribution theory. We first define both the $\mathcal{T}$-transform and the S-transform on the space (L^2). Later we extend them to the space of generalized white noise functionals (Hida distributions).

Definition 3.3 Let φ be (L^2)-functional, then $\mathcal{T}$-transform is defined by

$$(\mathcal{T}\varphi)(\xi) = \int_{E^*} \exp[i\langle x, \xi\rangle]\varphi(x)d\mu(x), \ \xi \in E. \tag{3.5.1}$$

Note that $\mathcal{T}$-transform is well defined since $|\exp i[\langle x, \xi\rangle]| = 1$, and φ is square integrable (hence integrable). This looks like a Fourier transform but not quite.

We then introduce a similar transformation that carries an (L^2)-functional $\varphi(x)$ also to a functional of ξ, where $\xi \in E$.

Definition 3.4 For $\varphi(x) \in (L^2)$, the S-transform is defined by

$$(S\varphi)(\xi) = C(\xi) \int_{E^*} \exp[\langle x, \xi\rangle]\varphi(x)d\mu(x), \ \xi \in E, \tag{3.5.2}$$

where $C(\xi) = \exp[-\frac{1}{2}\|\xi\|^2]$.

We note that S-transform is well defined too, since $\exp[\langle x, \xi\rangle]$ is μ-square integrable, that is, the transformation S is well defined on (L^2). In fact $\langle x, \xi\rangle$ is an $N(0, \|\xi\|^2)$-variable. Also, note that under the S-transform real $\varphi(x)$ goes to real functional of ξ.

The functional $(S\varphi)(\xi)$ is often called the *U-functional* associate with φ and is denoted by $U(\xi)$. That is,

$$(S\varphi)(\xi) = U(\xi).$$

Relationship between the two transforms $\mathcal{T}$ and S can be seen as follows.

Proposition 3.4 *If $\varphi \in \mathcal{H}_n$, then*

$$(\mathcal{T}\varphi)(\xi) = i^n C(\xi)U(\xi).$$

Proof.

$$\int e^{i\langle x,\xi\rangle} e^{t\langle x,f\rangle - \frac{1}{2}t^2\|f\|^2} d\mu = e^{-\frac{1}{2}t^2\|f\|^2} \int e^{\langle x, i\xi + tf\rangle} d\mu$$
$$= e^{-\frac{1}{2}\|\xi\|^2 + it\langle x, i\xi + t\rangle}$$

Thus, we have

$$\mathcal{T}(e^{t\langle x,f\rangle - \frac{1}{2}t^2\|f\|^2}) = C(\xi)e^{it\langle \xi,f\rangle}.$$

On the other hand

$$S(e^{t\langle x,f\rangle - \frac{1}{2}t^2\|f\|^2}) = e^{t\langle \xi,f\rangle}.$$

Accordingly, the assertion follows. ∎

We claim that the right-hand side of (3.5.2) is equal to

$$\int_{E^*} \varphi(x+\xi)d\mu(x),$$

which can be proved as follows.

For the computation we can use the translation formula (Hida[27], Section 5.7) for the Gaussian measure μ:

$$\frac{d\mu(\cdot - \xi)}{d\mu(\cdot)} = e^{\langle \cdot, \xi\rangle - \frac{1}{2}\|\xi\|^2}, \xi \in L^2(R).$$

Thus,

$$\begin{aligned}
\int_{E^*} \varphi(x+\xi)d\mu(x) &= \int_{E^*} \varphi(y)d\mu(y-\xi), \\
&= \int_{E^*} \varphi(y)e^{\langle y,\xi\rangle - \frac{1}{2}\|\xi\|^2} d\mu(y) \\
&= e^{-\frac{1}{2}\|\xi\|^2} \int_{E^*} e^{\langle y,\xi\rangle}\varphi(y)d\mu(y) \\
&= (S\varphi)(\xi).
\end{aligned}$$

Set

$$\mathcal{F} = \{S\varphi; \varphi \in (L^2)\}. \qquad (3.5.3)$$

Theorem 3.1 *The space $\mathcal{F}$ can be made being a Reproducing Kernel Hilbert Space (RKHS) with the reproducing kernel $C(\xi - \eta)$ such that*

$$S(L^2) \cong (L^2)$$

under the mapping

$$\varphi \mapsto S\varphi \quad \text{with } \varphi \in (L^2).$$

Outline of the proof

Consider the correspondence

$$S : e^{\langle x, \cdot \rangle} \longmapsto C(\cdot - f) = \int e^{\langle x, \cdot \rangle} e^{\langle x, f \rangle} d\mu(x)$$

and

$$\langle e^{-\langle x, f \rangle}, e^{\langle x, g \rangle} \rangle_{(L^2)} = E(e^{-\langle x, f \rangle} e^{-\langle x, g \rangle}) = C(g - f).$$

Hence

$$(C(\cdot - f), C(\cdot + g))_{\mathcal{F}} = C(g - f).$$

Since $\{e^{-\langle x, f \rangle}, f \in L^2(R^1)\}$ generate the space $\mathcal{F}$, the above equality implies

$$(F(\cdot), C(\cdot + g) = F(g)$$

which proves the reproducing property.

Thus, we conclude that the space $\mathcal{F}$, becomes a reproducing kernel Hilbert space with kernel $C(g - f)$, and we have

$$(L^2) \cong \mathcal{F}.$$

∎

Example 3.2 Let $\varphi(\dot{B}) = \int f(t) \dot{B}(t) dt$. Then,

$$(S\varphi)(\xi) = \int f(t) \xi(t) dt.$$

Example 3.3 Let $\varphi(\dot{B}) = \langle \dot{B}, \xi_1 \rangle \langle \dot{B}, \xi_2 \rangle$, where $\xi_1(t)\xi_2(t) = 0$ holds. Then $\langle \dot{B}, \xi_1 \rangle$ and $\langle \dot{B}, \xi_2 \rangle$ are independent. The S-transform of $\varphi(\dot{B})$ is

$$\begin{aligned} S(\varphi)(\xi) &= \langle \xi_1, \xi \rangle \langle \xi_2, \xi \rangle \\ &= \int \int F(u,v)\xi(u)\xi(v)dudv \end{aligned}$$

where $F(u,v) = \xi_1(u)\hat{\otimes}\xi_2(v)$, the symmetric tensor product of ξ_1 and ξ_2.

The computation of the S-transform of $\varphi(\dot{B})$ is as follows.

$$(S\varphi)(\xi) = e^{-\frac{1}{2}\|\xi\|^2} \int e^{\langle x, \xi \rangle} \langle x, \xi_1 \rangle \langle x, \xi_2 \rangle d\mu(x).$$

ξ is decomposed into three parts which are mutually orthogonal :

$$\xi = a\xi_1 + b\xi_2 + \xi^{\perp},$$

where $\xi^{\perp}$ is orthogonal to ξ_1 and ξ_2. Hence $\langle x, \xi \rangle$ is decomposed into three mutually independent random variables :

$$\langle x, \xi \rangle = a\langle x, \xi_1 \rangle + b\langle x, \xi_2 \rangle + \langle x, \xi^{\perp} \rangle.$$

The S-transform is expressed in the form

$$\begin{aligned} (S\varphi)(\xi) &= e^{-\frac{1}{2}\|\xi\|^2} \int e^{a\langle x, \xi_1 \rangle + b\langle x, \xi_2 \rangle + \langle x, \xi^{\perp} \rangle} \langle x, \xi_1 \rangle \langle x, \xi_2 \rangle d\mu(x) \\ &= e^{-\frac{1}{2}\|\xi\|^2} \int e^{a\langle x, \xi_1 \rangle} \langle x, \xi_1 \rangle d\mu(x) \int e^{b\langle x, \xi_2 \rangle} \langle x, \xi_2 \rangle d\mu(x) \int e^{\langle x, \xi^{\perp} \rangle} d\mu(x) \\ &= e^{-\frac{1}{2}\|\xi\|^2} e^{a^2 \frac{1}{2}\|\xi\|^2} a\langle \xi_1, \xi_1 \rangle e^{b^2 \frac{1}{2}\|\xi_2\|^2} b\langle \xi_2, \xi_2 \rangle e^{\frac{1}{2}\|\xi^{\perp}\|^2} \\ &= a\|\xi_1\|^2 \cdot b\|\xi_2\|^2. \end{aligned} \tag{3.5.4}$$

Coming back to the decomposition $\xi = a\xi_1 + b\xi_2 + \xi^{\perp}$, we know

$$\langle \xi, \xi_1 \rangle = a\|\xi_1\|^2, \ \langle \xi, \xi_2 \rangle = b\|\xi_2\|^2,$$

so that the last term of the above equation (3.5.4) is equal to

$$\langle \xi, \xi_1 \rangle \cdot \langle \xi_1, \xi_2 \rangle.$$

Or equivalently

$$S(\varphi)(\xi) = \int \int F(u,v)\xi(u)\xi(v)dudv.$$

Example 3.4 Exponential function.

Let η be a member of E. Define

$$\varphi(x) = \exp[\langle x, \eta \rangle] - \frac{t^2}{2}\|\eta\|^2 \quad t \in R^1.$$

Then,

$$(S\varphi)(\xi) = \exp[t\langle \eta, \xi \rangle].$$

Remark 3.1 *A white noise functional φ is expressed as a (nonlinear, in general) functional of a generalized function x, that is, $\varphi(x), x \in E^*$, so that it usually does not have a simply visualized expression. A better expression would be expected through the S-transform which is an analogue of the Laplace transform, at least in expression. Taking the S-transform of white noise functional, we obtain a functional of a C^∞-function ξ, thus, we are ready to appeal to the established theory of functional analysis. Similar story holds for the $\mathcal{T}$-transform.*

Lemma 3.1 *The following equality holds.*

$$S\left(H_k(\frac{\langle x, \eta \rangle}{\sqrt{2}}\right)(\xi) = 2^{\frac{k}{2}} (\eta, \xi)^k.$$

Proof. Take $H_k(\frac{\langle x,\eta \rangle}{\sqrt{2}})$ with $\|\eta\| = 1$. Then the S-transform is

$$\begin{aligned}
U(\xi) &= S\left(H_k(\frac{\langle x, \eta \rangle}{\sqrt{2}})\right) \\
&= C(\xi) \int e^{\langle x, \xi \rangle} H_k(\frac{\langle x, \eta \rangle}{\sqrt{2}}) d\mu(x) \\
&= C(\xi) \int e^{\langle x, a\eta + \xi^\perp \rangle} H_k(\frac{\langle x, \eta \rangle}{\sqrt{2}}) d\mu(x),
\end{aligned}$$

where $\xi = a\eta + \xi^\perp$, $\xi^\perp$ being orthogonal to η, so that $\|\xi\|^2 = a^2 + \|\xi^\perp\|^2$. Since $\langle x, \eta \rangle$ and $\langle x, \xi^\perp \rangle$ are independent,

$$\begin{aligned}
U(\xi) &= C(\xi) \int e^{\langle x, \xi^\perp \rangle} d\mu(x) \int e^{a\langle x, \eta \rangle} H_k(\frac{\langle x, \eta \rangle}{\sqrt{2}}) d\mu(x) \\
&= C(\xi) e^{\frac{1}{2}\|\xi^\perp\|^2} \int e^{ay} H_k(\frac{y}{\sqrt{2}}) d\mu(x), \quad \text{where } y = \langle x, \eta \rangle \\
&= C(\xi) e^{\frac{1}{2}\|\xi^\perp\|^2} a^k 2^{\frac{k}{2}} e^{\frac{a^2}{2}}
\end{aligned}$$

$$= 2^{\frac{k}{2}} \langle \xi, \eta \rangle^k.$$

■

Theorem 3.2 *The S-transform gives an isomorphism*

$$\mathcal{H}_n \cong \sqrt{n!}\widehat{L^2(R^n)}.$$

The symbol 'hat' on the right-hand side means the symmetric L^2 space.

Proof. Take a complete orthonormal system $\{\eta_k\}$ in $L^2(R^1)$ such that $\eta_k \in E$ for every k. Then, $\xi = \sum_k a_k \eta_k$, and we have

$$\langle x, \xi \rangle = \langle x, \sum_k a_k \eta_k \rangle.$$

Any $\varphi \in \mathcal{H}_n$ can be expressed as a linear functional of (in fact, the Fourier-Hermite polynomials in) the $\langle x, \eta_n \rangle$'s of the form

$$\begin{aligned} \varphi(x) &= \varphi_{\{n_k\}}(x) \\ &= \prod_{\sum n_k = n} H_{n_k}\left(\frac{\langle x, \eta_k \rangle}{\sqrt{2}}\right). \end{aligned}$$

Now take the S-transform to have

$$\begin{aligned} (S\varphi_{\{n_k\}})(\xi) &= C(\xi) \int \exp(\langle x, \xi \rangle) \left(\prod_k H_{n_k}\left(\frac{\langle x, \eta_k \rangle}{\sqrt{2}}\right) \right) d\mu(x), \ \sum n_k = n \\ &= C(\xi) \int \exp\left(\langle x, \sum_{j \neq n_k} a_j \eta_j \rangle + \langle x, \sum_{j = n_k} a_j \eta_j \rangle \right) \cdot \\ &\qquad \left(\prod_k H_{n_k}\left(\frac{\langle x, \eta_k \rangle}{\sqrt{2}}\right) \right) d\mu(x) \\ &= C(\xi) \left(\int \exp(\langle x, \sum_{j \neq n_k} a_j \eta_j \rangle) d\mu(x) \right) \cdot \\ &\qquad \left(\int \exp(\langle x, \sum_{j = n_k} a_j \eta_j \rangle) \prod H_{n_k}\left(\frac{\langle x, \eta_k \rangle}{\sqrt{2}}\right) d\mu(x) \right) \\ &= C(\xi) \left(\prod_{j \neq n_k} \int \exp(\langle x, a_j \eta_j \rangle) d\mu(x) \right) \left(\prod_k S(H_{n_k}\left(\frac{\langle x, \eta_k \rangle}{\sqrt{2}}\right))(a_k \eta_k) \right) \end{aligned}$$

$$= C(\xi)\left(\prod_{j\neq n_k}\exp(\frac{1}{2}a_j^2)\right)\left(\prod_{j=n_k}\exp(\frac{1}{2}a_j^2)\right)\left(\prod_k 2^{\frac{n_k}{2}}\langle\eta_k, a_k\eta_k\rangle^{n_k}\right)$$

$$= C(\xi)\left(\exp(\frac{1}{2}\sum_j a_j^2)\right)\left(\prod_k 2^{\frac{n_k}{2}}a_k^{n_k}\right)$$

$$= 2^{\frac{n}{2}}\prod_k\langle\xi,\eta_k\rangle^{n_k}$$

$$= \int\cdots\int F(t_1,\cdots t_n)\xi(t_1)\cdots\xi(t_n)dt_1\cdots dt_n,$$

where

$$F(t_1,\cdots,t_n) = 2^{\frac{n}{2}}\prod_{\sum n_k=n}\eta^{\otimes n_k}.$$

It is seen that the value of the integral is not changed if F is replaced by its symmetrization

$$\widehat{F}(t_1,\cdots,t_n) = \frac{1}{n!}\sum F(t_1,\cdots,t_n)$$

where the sum extends over all permutations of $\{t_1,t_2,\cdots,t_n\}$.
Thus, we obtain

$$(S\varphi^{(n)})(\xi) = \int\cdots\int\widehat{F}(t_1,\cdots t_n)\xi(t_1)\cdots\xi(t_n)dt_1\cdots dt_n. \qquad (3.5.5)$$

We have

$$\|F\| = 2^{\frac{n}{2}}$$

so that

$$\|\hat{F}\| = \left(\prod_{\sum n_k=n} n_k!\right)^{\frac{1}{2}}(n!)^{\frac{1}{2}}2^{\frac{n}{2}}.$$

$$\int_{\mathcal{H}_n}|\varphi^{(n)}(x)|^2d\mu(x) = \int_{\mathcal{H}_n}\prod_{\sum n_k=n}H_{n_k}(\frac{\langle x,\eta_k\rangle}{\sqrt{2}})^2d\mu(x)$$

$$= \prod_{\sum n_k=n}(2\pi)^{-\frac{1}{2}}\int_{\mathcal{H}_n}H_{n_k}(\frac{t}{\sqrt{2}})e^{-\frac{1}{2}t^2}dt$$

$$= \prod_{\sum n_k = n} n_k! 2^{n_k}$$
$$= 2^n \prod_{\sum n_k = n} n_k!.$$

That is,

$$\|\varphi^{(n)}\|_{(L^2)} = 2^{\frac{n}{2}} \left(\prod_{\sum n_k = n} n_k! \right)^{\frac{1}{2}}.$$

Thus we have

$$\|\varphi^{(n)}\|_{(L^2)} = \sqrt{n!}\|\hat{F}\|_{L^2(R^n)}.$$

Hence, the assertion follows. ■

Because of the expression (3.5.5), the theorem gives an *integral representation* of a white noise functional. Such a functional is uniquely determined in any space $\mathcal{H}_n$.

That is, we have, for $\varphi^{(n)} \in \mathcal{H}_n$,

$$(S\varphi^{(n)})(\xi) = \int_{R^n} F^{(n)}(u)\xi^{\otimes n}(u)du^n,$$

where $F^{(n)} \in \widehat{L^2(R^n)}$ which is a symmetrization of $L^2(R^n)$.

Finally, we can assert that S-transform gives the integral representation of any (L^2)-functional in terms of a series of symmetric $L^2(R^n)$-functions.

Example 3.5 (Hida[24]) Real quadratic functionals

Let $\varphi(\dot{B})$ be a real-valued functional in $\mathcal{H}_2$. Then, its S-transform is of the form

$$(S\varphi)(\xi) = \int\int F(u,v)\xi(u)\xi(v)dudv,$$

where $F(u,v)$ is a symmetric real valued function in $L^2(R^2)$.

Assume that

$$F(u,v) = 0, \quad (u,v) \notin [-T,T]^2, T > 0.$$

The kernel $F(u, v)$ defines an integral operator acting on $L^2(R^1)$ and it is of Hilbert-Schmidt type. We have the eigensystem $\{\lambda_n, \eta_n, n \geq 0\}$. That is

$$\lambda_n \int F(u, v)\eta_n(v) = \eta_n(u).$$

We, therefore, have an expression

$$\varphi(\dot{B}) = \sum \lambda_n^{-1}(\dot{B}(\eta_n)^2 - 1)$$

since $F(u, v)$ is expressed in the form

$$F(u, v) = \sum \lambda_n^{-1}\eta_n(u)\eta_n(v), \quad \sum \lambda_n^{-2} < \infty.$$

By using this expression of $\varphi(\dot{B})$ we can show that the characteristic function $\psi(z)$ of $\varphi(\dot{B})$ is

$$\psi(z) = E(e^{iz\varphi(\dot{B})}) = \prod_n (1 - 2iz\lambda_n^{-1})^{-\frac{1}{2}} \exp[-iz\lambda_n^{-1}], \ z : \text{real}.$$

In terms of the modified Fredholm determinant $\delta(z, F)$ (see Smithies[126]) we have

$$\psi(z) = \delta(2iz, F)^{-1}.$$

As a result, we can claim that the n-th order semi-invariant γ_n of $\varphi(\dot{B})$ is given by

$$\begin{cases} \gamma_1 = 0 \\ \gamma_n = (-2)^{n-1}(n-1)! \sum_k \lambda_k^{-n}, \qquad n \geq 2. \end{cases}$$

Moreover, we can see an interesting formulas on semi-invariant in the particular case where

$$F(-u, -v) = -F(u, v).$$

Namely,

$$\begin{cases} \gamma_{2n+1} = 0 \\ \gamma_{2n} = 2^{2n}(2n-1)! \sum_k \lambda_k^{-2n}, \qquad n \geq 1. \end{cases}$$

Example 3.6 Stochastic area

Following P. Lévy (proposed in Lévy[87]) introduced a stochastic area $S(T)$ which defines the area of the region enclosed by the curve of a two-dimensional Brownian motion $(B_1(t), B_2(t)), 0 \le t \le T$, and the chord connecting the origin with the terminal point $(B_1(t), B_2(t))$. In general, the area enclosed by a smooth curve $(x(t), y(t))$ is defined by

$$\frac{1}{2}\int_0^T (x(t)y'(t) - x'(t)y(t))dt.$$

A Brownian curve is not smooth, but we can imitate this integral by taking a stochastic integral (see Chapter 5).

Namely, we define $S(T)$ by

$$S(T) = \frac{1}{2}\int_0^T (B_1(t)dB_2(t) - B_2(t)dB_1(t)).$$

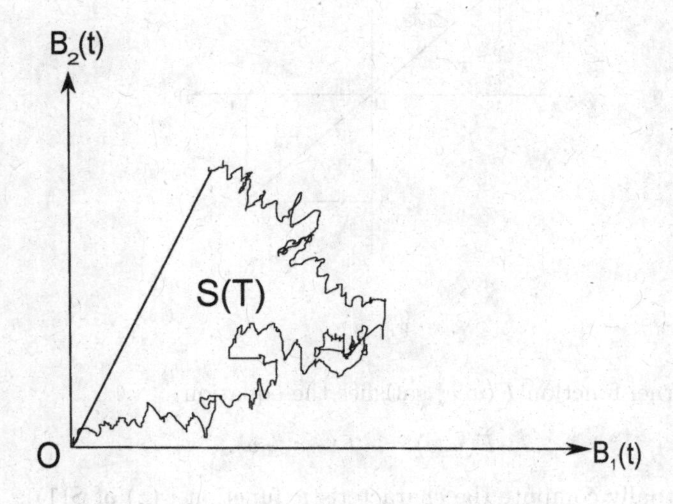

Fig. 3.1

Since $B_1(t)$ and $B_2(t)$ are independent, this integral is well defined. The analysis on this topic is carried on as follows. Take a Brownian motion $B(t)$ and define

$$B_1(t) = B(t),$$
$$B_2(t) = B(-t),$$

for $t \geq 0$. Define

$$S(T) = \frac{1}{2}\int_0^T (B(t)(-\dot{B}(-t)) - B(-t)\dot{B}(t))dt.$$

Then, $S(T)$ is a member of $\mathcal{H}_2$. The kernel function $F(u,v)$ of its S-transform is

$$F(u,v) = \begin{cases} -\frac{1}{4} & \text{if } uv \leq 0, u, v \leq T, -v < u, \\ \frac{1}{4} & \text{if } uv \leq 0, u, v \geq T, -v > u, \\ 0 & \text{otherwise.} \end{cases}$$

Taking $T = 1, F$ is illustrated by Fig. 3.2.

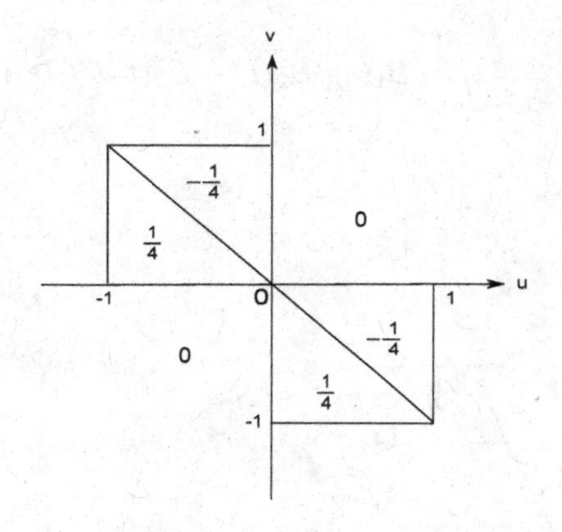

Fig. 3.2

The kernel function $F(u,v)$ satisfies the equation

$$F(u,v) = -F(-v,-u).$$

We can actually compute the characteristic function $\psi(z)$ of $S(1)$:

$$\psi(z) = \{cosh(\frac{z}{2})\}^{-1}.$$

Thus, we are given *Meixner* distribution.

The semi-invariants are

$$\begin{cases} \gamma_{2p+1} = 0, & p \geq 0 \\ \gamma_{2p} = (2^{2p} - 1)!\frac{B_p}{4p}, & p \geq 1, \end{cases}$$

where B_p denotes the p-th Bernoulli number.

It is very interesting to study the way of time-development of the stochastic area.

Using the representation of the kernel $F(t; u, v)$ associated with $S(t)$ it is easy to see :

Proposition 3.5 $S(t), t \geq 0$, *has orthogonal increments.*

Then we express the time development of $S(t)$ in the following manner.

Proposition 3.6 *Dilation* $t \mapsto at, a > 0$, *applied to* $B(t)$ *gives the same process as* $aS(t)$ *which indicates the self-similarity of the stochastic area.*

3.6 White noise $\dot{B}(t)$ related to δ-function

The idea behind the $\mathcal{T}$-transform is **factorization**. We are going to explain this fact by taking a particular and important example.

We might say that white noise $\dot{B}(t)$ is a random square root of the δ-function:

$$(\text{stochastic})\sqrt{\delta_t} = \dot{B}(t). \tag{3.6.1}$$

The expression is, of course, formal, but seems to be reasonable in a sense that

$$E(\dot{B}(t)\dot{B}(s)) = \delta(t-s),$$

and a correct interpretation will be given below.

It is noted that $\dot{B}(t)$ is *atomic* as an *idealized elemental* random variable. While Brownian motion $\{B(t), t \in R^1\}$ is atomic as a stochastic process, where the *causality* should be taken into account.

As is noted before, in white noise analysis, the $\mathcal{T}$-transform and the S-transform play dominant roles both explicitly and implicitly. As one of their roles, we can see factorization problem for positive definite functions.

We now come to the *factorization problem.* We first give a concrete example of factorization of covariance function. We refer to the literature[70] for the Karhunen theory.

For the moving average representation of a weakly stationary stochastic process $X(t)$ with $E(X(t)) = 0$, is of the form:

$$X(t) = \int_{-\infty}^{t} F(t-u)dZ(u), \tag{3.6.2}$$

where $Z(u)$ is a process with stationary orthogonal increments such that $E(|dZ(u)|^2) = du$. We expect that the representation is canonical, that is, the linear space $M_t(X)$ and $M_t(Z)$ is the same for every t, where $M_t(X)$ is spanned by $X(s), s \le t$.

Determination of the canonical kernel $F(t, u)$ (Hida[21]) is discussed in line with the factorization of the covariance function

$$\gamma(h) = E(X(t+h)X(t)) \tag{3.6.3}$$

in such a way that the following equality holds:

$$\gamma(h) = \int^{t} F(t+h-u)F(t-u)du. \tag{3.6.4}$$

Note. The canonical representation theory is recalled briefly in Chapter 7.

Remark 3.2 *This theory is applied to the problem of obtaining the canonical representation of a Gaussian process when the process is stationary, mean continuous and purely deterministic.*

Remark 3.3 *The kernel F (indeed, the optimal kernel) is obtained from the factorization of the covariance function (3.6.4) explicitly, where the method is purely analytic.*

Coming back to the problem on $\dot{B}(t)$, we may say its covariance function is the delta function. The product $\dot{B}(t)\cdot\dot{B}(s)$ implies a δ-function by taking expectation. *This is the reason why $\dot{B}(t)$ is said to be the stochastic square root of a δ-function.*

Addendum : Reproducing Kernel Hilbert Space and $\mathcal{T}$-transform

Given a positive definite function $\Gamma(t, s), t, s \in T$. Then, there is a linear space $\boldsymbol{F}_1$ spanned by $\Gamma(\cdot, t), t \in T$, where we have a bilinear form

$$(f(\cdot), \Gamma(\cdot, s)) = f(s) \quad \text{(reproducing property).}$$

This equation defines a semi-norm $\|\cdot\|$ in $\boldsymbol{F}_1$. We consider such a minimal space and define a factor space $\boldsymbol{F} = \boldsymbol{F}_1/\|\cdot\|$. Thus, we obtained $\boldsymbol{F}$ which is a Hilbert space where the reproducing property holds. The kernel $\Gamma(t,s)$ is called the *reproducing kernel* and $\boldsymbol{F}$ is the *reproducing kernel Hilbert space* (abbr, R.K.H.S.). To specify the kernel, the above $\boldsymbol{F}$ is often written as $\boldsymbol{F}(\Gamma)$.

We have, in particular

$$(\Gamma(\cdot,t), \Gamma(\cdot,s)) = \Gamma(s,t), \tag{3.6.5}$$

where one can see a square root (in a sense) of the positive definite function Γ or its factorization.

There are two directions in the application to the theory of stochastic processes.

1) Covariance function and linear theory.

Consider a stochastic process $X(t), t \in T$. Assume that $E(X(t)) = 0$ and $E(|X(t)|^2) < \infty$.

To fix the idea, let us consider a Gaussian process $X(t), t \in R^1$ with mean 0 and with covariance function $\Gamma(t,s)$.

Associated with Γ is a R.K.H.S. $\boldsymbol{F}(\Gamma)$, where $X(t)$ corresponds to $\Gamma(\cdot,t) \in \boldsymbol{F}(\Gamma)$ which is shown by (3.6.5) (the factorization of the covariance function). Then, we are led to have the canonical kernel, although we do not go into details to this direction. (cf. Hida[21].)

2) Application to nonlinear theory.

We now come to a *nonlinear* theory, where nonlinear functions of a stochastic process are dealt with. The present case takes care of nonlinear functionals of white noise $\dot{B}(t)$'s or those of $x \in E^*(\mu)$, μ being the white noise measure, i.e. the probability distribution of white noise $\dot{B}(t), t \in R^1$.

The characteristic functional $C(\xi) = \exp[-\frac{1}{2}\|\xi\|^2], \xi \in E$ of white noise defines a R.K.H.S. $\boldsymbol{F}(C)$, so that we can play the same game as in 1).

We then consider the complex Hilbert space $(L^2) = L^2(E^*, \mu)$ involving nonlinear functionals of the smeared functions of $\dot{B}(t)$ or of $x \in (E^*, \mu)$ with finite variance. This space is classical. It is generated by the $e^{ia\langle x,\xi\rangle}, \xi \in$

$E, a \in R^1$. The $X(t)$ in (3.6.3) is replaced by $e^{i\langle x,\xi\rangle}$, so that the covariance is now

$$C(\xi - \eta) = E(e^{i\langle x,\xi\rangle} e^{-i\langle x,\eta\rangle}), \tag{3.6.6}$$

where $C(\xi)$ is the characteristic functional of white noise. It is positive definite, so that we can form a reproducing kernel Hilbert space $\boldsymbol{F}$ with the reproducing kernel $C(\xi - \eta)$ where C is the characteristic functional of white noise.

Observe now the formula (rephrasement of (3.6.6)).

$$C(\xi - \eta) = \int e^{i\langle x,\xi\rangle} \overline{e^{i\langle x,\eta\rangle}} d\mu(x). \tag{3.6.7}$$

The factor $\overline{e^{i\langle x,\eta\rangle}}$ of the product of the integrand can be extended to a general white noise functional $\varphi(x)$. Then, the integral turns into the following formula

$$(\mathcal{T}\varphi)(\xi) = \int e^{i\langle x,\xi\rangle} \varphi(x) d\mu(x), \tag{3.6.8}$$

which is the $\mathcal{T}$-transform of $\varphi(x)$. Let it be denoted by $V_\varphi(\xi)$ or simply by $V(\xi)$. It holds that

$$(V(\cdot), C(\cdot - \xi)) = V(\xi), \tag{3.6.9}$$

where $(\cdot, \cdot)$ is the inner product in the R.K.H.S. $\boldsymbol{F}$, in particular,

$$(C(\cdot - \eta), C(\cdot - \xi)) = C(\xi - \eta), \tag{3.6.10}$$

which is, in other words, the factorization of $C(\xi - \eta)$. Note that $C(\xi - \eta) = C(\eta - \xi)$ in the present case.

Thus the characteristic functional is factorized by the $\mathcal{T}$-transform with the help of the theory of R.K.H.S.

Theorem 3.3 *The following facts hold.*

i) The $\mathcal{T}$-transform factorizes the characteristic functional $C(\xi)$ of white noise.

ii) The system $\{C(\cdot - \xi)\}$ corresponds to the system $\{e^{ia\langle x.\xi\rangle}\}$ which is total in (L^2).

The proofs of i) and ii) have already been given above, as for the proof we need some interpretations which can be seen in [6].

Note. It is very important to recognize the intrinsic meaning of the $\mathcal{T}$ and the S-transforms including the facts mentioned above. Needless to say, the topology equipped with R.K.H.S., the image of the $\mathcal{T}$ or S-transform, is convenient for white noise analysis. For one thing, all the members in those spaces are functionals of smooth functions, and for another thing, the topology is pointwise convergence, so that we have visualized expressions. Those transforms should **never** be thought of (simply) as similar transforms to some classical transforms, which are automorphisms within the given space.

What we shall observe in what follows is not a rigorous assertion, but it would be acceptable as an intuitive interpretation. As was briefly mentioned "Stochastic square root, as it were, of the Dirac delta function."

We shall give a naive verification on the problem in question. Let us state a proposed and formal assertion.

Assertion. The delta function δ_t is positive definite.

Proof. For any n and for any complex numbers $z_j, 1 \leq j \leq n$, we have

$$\sum_{j,k} z_j \bar{z_k} \delta(t_j - t_k) = \left(\sum_j z_j \delta(\cdot - t_j), \sum_k z_k \delta(\cdot - t_k) \right)$$
$$= \sum_j |z_j|^2 \delta_j(0) \geq 0.$$

(End of Addendum)

We now give an interpretation on what we wish to claim.

Factorization problem in terms of R.K.H.S.

We start with the δ-function $\delta_t(\cdot) = \delta(\cdot - t)$. It is a generalized function in $K^{-1}(R^1)$ and its Fourier transform is $\frac{e^{it\lambda}}{\sqrt{2\pi}}$. Define a mapping Π:

$$\Pi : \delta_t \;\mapsto\; e^{it\lambda}.$$

The mapping Π defines a bijection between two systems:

$$\Pi : \Delta = \{\delta_t, t \in R^1\} \;\mapsto\; \Lambda = \{e^{it\lambda}, t \in R^1\}.$$

The $K^{-1}(R^1)$-norm is introduced to Δ, so is topologized Λ. Their closures are denoted by the same symbols, respectively. And they are isomorphic to each other. The inner product of $e^{it\lambda}$ and $e^{is\lambda}$ in $K^{-1}(R^1)$ is

$$\int \frac{e^{i(t-s)\lambda}}{\pi(1+\lambda^2)} d\lambda = e^{-|t-s|}, \tag{3.6.11}$$

which is positive definite. (In fact, it is the covariance function of Ornstein-Uhlenbeck's Brownian motion.) Hence, we can form a R.K.H.S. $\boldsymbol{F}_\delta$ with reproducing kernel $e^{-|t-s|}$. We can establish a mapping through Π:

$$\delta(\cdot - t) \quad \mapsto \quad e^{-|\cdot - t|}. \tag{3.6.12}$$

The inner product

$$\langle e^{-|\cdot - t|}, e^{-|\cdot - s|} \rangle = e^{-|t-s|}$$

implies

$$\langle \delta(\cdot - t), \delta(\cdot - s \rangle_\Delta = \delta(t-s). \tag{3.6.13}$$

Note that the mapping $S(\dot{B}(t))(\xi) = \xi(t) = \int \delta(t-u)\xi(u)du$, thus it is seen that $\dot{B}(t)$ corresponds to $\delta(t-u)$. Now the right-hand side of (3.6.13) is just the delta function and the left-hand side is viewed as the (inner) *product* of $\dot{B}(t)$ and $\dot{B}(s)$.

Summing up we have

Proposition 3.7 *In view of equation (3.6.13), a stochastic square root of the delta function is a white noise.*

Observations

We now make again an important *remark* on the $\mathcal{T}$- and the $\mathcal{S}$-transforms regarding their meanings and roles, through various observations which are in order.

1) They define mappings from the space of white noise functionals to the spaces with reproducing kernels. We are given big advantages, since the image involves good functionals of smooth function ξ. Those functionals are no longer random, and are easy to be analyzed, in general, by appealing to the known theory of functional analysis.

2) They help us to determine factorizations of covariances and others.

3) They play a role of determining the integral representations of white noise functionals. Together with the use of the Sobolev spaces we have been led to introduce spaces of *generalized* white noise functionals. To be more concrete, we can form, for instance, the space $\mathcal{H}_n^{(-n)}$ of the space of generalized white noise functionals of degree n as an extension of $\mathcal{H}_n$. (See the next section.)

3.7 Infinite dimensional space generated by Hermite polynomials in $\dot{B}(t)$'s of higher degree

We now introduce $\mathcal{H}_n^{(-n)}$, the space of generalized white noise functionals of degree n. According to Theorem 3.2, we know that there is an isomorphism between $\mathcal{H}_n$ and $\widehat{L^2(R^n)}$. We note that $\mathcal{H}_n$ is spanned by the Hermite polynomials in $\langle \dot{B}, \xi_n \rangle$'s of degree n. We also know that there is a mapping Π_n which defines an isomorphism

$$\Pi_n : \mathcal{H}_n \cong \sqrt{n!}\widehat{L^2(R^n)},$$

where $\mathcal{H}_n$ is the space of homogeneous chaos of degree n. Let $K^m(R^n)$ be the Sobolev space of order m over R^n. The subspace of $K^m(R^n), m \in \mathbf{Z}, n \geq 1$, involving symmetric functions will be denoted by $\widehat{K^m(R^n)}$. Take the dual space $\widehat{K^{-m}(R^n)}$ of $\widehat{K^m(R^n)}$ with respect to the topology being equipped with $L^2(R^n)$. Then, we have a Gel'fand triple

$$\widehat{K^m(R^n)} \subset \widehat{L^2(R^n)} \subset \widehat{K^{-m}(R^n)}, \tag{3.7.1}$$

where the inclusions are continuous injections.

If we take $m = \frac{n+1}{2}$, it is convenient to carry out the analysis on $\mathcal{H}_n$ since

i) when the members in $\widehat{K^{\frac{n+1}{2}}(R^n)}$ are restricted to an $(n-1)$-dimensional hyperplane, they belong to $\widehat{K^{\frac{n}{2}}(R^{n-1})}$, i.e. the degree decreases by $\frac{1}{2}$ when the restriction is done to a space of one-dimensional lower, and

ii) if the integral of a member in $\widehat{K^{-\frac{n+1}{2}}(R^n)}$ is taken on the diagonal, the degree decreases by one,

which follow from the trace theorem[91] in the theory of Sobolev space.

Thus taking $m = \frac{n+1}{2}$, we have the triple

$$\widehat{K^{\frac{n+1}{2}}(R^n)} \subset \widehat{L^2(R^n)} \subset \widehat{K^{-(\frac{n+1}{2})}(R^n)} \tag{3.7.2}$$

and we form the triple

$$\mathcal{H}_n^{(n)} \subset \mathcal{H}_n \subset \mathcal{H}_n^{(-n)}, \tag{3.7.3}$$

where each space of the triple is isomorphic to the corresponding symmetric Sobolev space. The inclusions from left to right are continuous injections.

The dual space $\mathcal{H}_n^{(-n)}$ of $\mathcal{H}_n^{(n)}$ is defined in the usual manner, based on the scalar product in $\mathcal{H}_n$. The norms in these spaces are denoted by $\|\cdot\|_{-n}$ and $\|\cdot\|_n$, respectively and defined by the corresponding Sobolev norms. Namely, for any $n \in \boldsymbol{Z}$, if

$$\varphi(\in H_n^{(n)}) \longleftrightarrow f(\in \widehat{K^{\frac{n+1}{2}}})$$

then

$$\|\varphi\|_n = \|f\|_{\widehat{K^{\frac{n+1}{2}}}}.$$

Proposition 3.8 *The following isometric bijection is established*

$$\mathcal{H}_n^{(-n)} \cong \sqrt{n!}\widehat{K^{-(n+1)/2}(R^n)} \quad \text{(symmetric)},$$

for every $n \geq 1$.

Definition 3.5 The Hilbert space $\mathcal{H}_n^{(-n)}, n \geq 0$, is called a space of generalized functionals of degree n, on the space of Hida distributions of degree n.

3.8 Generalized white noise functionals

There are two typical methods of introducing the space of generalized white noise functionals, $(L^2)^-$ and $(S)^*$. Both are defined by the Gel'fand triple starting from $(L^2) = L^2(E^*, \mu)$ such that

$$(L^2)^+ \subset (L^2) \subset (L^2)^-$$

and

$$(S) \subset (L^2) \subset (S)^*.$$

Members in $(L^2)^-$ are called *generalized white noise functionals* and often called *Hida distributions* (called after Professor L. Streit), the same for those in the space $(S)^*$.

Here we wish to mention that the space of generalized white noise functionals $(L^2)^-$ is the Hilbert space equipped with norm and can be expressed as (3.8.2). By $\mathcal{H}_n^{(-n)}$'s, functionals are graded by n, the creation and the annihilation operators make the grade up and down, respectively. While, (S) is an algebra and complete metric space. For the members of $(S)^*$, Potthoff-Streit characterization is given.

Details will be given in what follows.

I. The space $(L^2)^-$

In Section 3.6, we have obtained the triples (3.7.2) and (3.7.3).

Denote the $\mathcal{H}_n^{(n)}$-norm by $\|\cdot\|_n$. Define

$$\|\cdot\|_+ = \Big(\sum_n c_n^2 \|\cdot\|_n^2\Big)^{\frac{1}{2}},$$

where c_n is a non-decreasing sequence of positive numbers which are suitably chosen according to the problem to be discussed.

The completion of the algebraic sum of $\mathcal{H}_n^{(n)}$ is denoted by $(L^2)^+$ with the norm $\|\cdot\|_+$. That is

$$(L^2)^+ = \bigoplus_n c_n \mathcal{H}_n^{(n)} \tag{3.8.1}$$

is the Hilbert space which plays the role of *test functionals.*

Definition 3.6 The dual space of $(L^2)^+$ with respect to the (L^2)-norm is a space of generalized white noise functionals or Hida distributions and is denoted by $(L^2)^-$.

By definition, the space $(L^2)^-$ is expressed as the direct sum of the form

$$(L^2)^- = \bigoplus_{n=0}^{\infty} c_n^{-1} \mathcal{H}_n^{(-n)} \tag{3.8.2}$$

and it is a Hilbert space too.

The norm introduced to $(L^2)^-$ is $(\sum_n c_n^{-2}\|\cdot\|_{-n}^2)^{\frac{1}{2}}$, where $\|\cdot\|_{-n}$ is the $\mathcal{H}_n^{-n}$-norm.

Hence, we obtain the triple

$$(L^2)^+ \subset (L^2) \subset (L^2)^-. \tag{3.8.3}$$

The canonical bilinear form connecting $(L^2)^+$ and $(L^2)^-$ is denoted by $\langle \cdot, \cdot \rangle_\mu$.

Proposition 3.9 *The algebra A^* generated by $\{: e^{a\dot{B}(t)} :; a \in \mathbf{C}, t \in R^1\}$ is dense in $(L^2)^-$.*

Proof. The generalized functional $: e^{a\dot{B}(t)} := e^{a\dot{B}(t) - \frac{1}{2}a^2(1/dt)}$ is the generating functions of Hermite polynomials. Finite products of such generating functions with different t_j's can generate all the Hermite polynomials in $\dot{B}(t_j)$'s. Hence, the assertion follows by the definition of $(L^2)^-$. Needless to say, the topology of convergence is that of $(L^2)^-$.

■

Note. One can view the topology introduced to the space $(L^2)^-$ to see the significance and suitability of the Sobolev norm (space).

Example 3.7 The functional $\varphi(\dot{B}(t)) = \sum H_n(\dot{B}(t); \frac{1}{dt})$ is a $(L^2)^-$-functional.

II. The space $(S)^*$

Before constructing the space $(S)^*$, we first recall the Schwartz distribution space on R^1.

Let $S(R^1)$ be the space of all real-valued rapidly decreasing functions f on R^1. That is for $f \in S(R^1)$, and we assume that $f \in C^\infty$ and

$$\lim_{n \to \infty} |x^n f^{(k)}(x)| = 0, \quad n, k \geq 0.$$

Define the norms $\|\cdot\|_{n,k}$ on $S(R^1)$ such that

$$\|f\|_{n,k} = \sup_{x \in R} |x^n f^{(k)}(x)|, \quad n, k \geq 0.$$

These norms are Hilbertian and consistent. The space $S(R^1)$, having been topologized by these norms, is a topological space which is proved to be nuclear. The dual space of $S(R^1)$ is denoted by $S'(R^1)$. Thus, we have a Gel'fand triple

$$S(R^1) \subset L^2(R^1) \subset S'(R^1). \tag{3.8.4}$$

We can give some interpretation on this triple as follows.

Take the operator

$$A = -\frac{d^2}{du^2} + u^2 + 1$$

on $L^2(R^1)$. Let e_n be defined by

$$e_n = \frac{1}{(\sqrt{\pi}2^n n!)^{-\frac{1}{2}}} H_n(x) e^{-\frac{x^2}{2}},$$

which is in $S(R^1)$. The family $\{e_n, n \geq 0\}$, is an orthonormal basis for $L^2(R^1)$. We can see that

$$Ae_n = (2n+2)e_n, \; n \geq 0.$$

Define

$$\|f\|_p = \|A^p f\|, \; p \in R.$$

Thus,

$$\|f\|_p = \left(\sum_{n=0}^{\infty} (2n+2)^{2p} (f, e_n)^2 \right)^{\frac{1}{2}},$$

where $(\cdot,\cdot)$ is the inner product in $L^2(R^1)$.

Let

$$S_p(R^1) = \left\{ f \in L^2(R), \; \|f\|_p < \infty \right\}, \quad p \geq 0, \tag{3.8.5}$$

where $S_0 = L^2(R^1)$.

The dual space S'_p of S_p can be identified as $S_{-p}(R^1)$ where $S_{-p}(R^1)$ is a completion of $(L^2(R^1), \|\cdot\|_{-p})$, by applying the Riesz representation theorem to $L^2(R^1)$, where $\|\cdot\|_{-p}$ is defined by

$$\|f\|_{-p} = \left(\sum_{n=0}^{\infty} (2n+2)^{-2p} (f, e_n)^2 \right)^{\frac{1}{2}}.$$

Thus, we have, for $p \geq 0$,

$$S_p(R^1) \subset L^2(R^1) \subset S_{-p}(R^1).$$

Take the projective and inductive limits

$$S(R^1) = \bigcap_{p \geq 0} S_p(R^1),$$
$$S'(R^1) = \bigcup_{p \geq 0} S_{-p}(R^1)$$

of $\{S_p(R^1), p \geq 0\}$ and $\{S_{-p}(R^1), p \geq 0\}$, respectively. Thus the triple (3.8.4) is obtained.

Generalized white noise functional is also defined by using the second quantization technique.

The linear operator A is densely defined on $L^2(R^1)$. Then, there is an operator $\Gamma(A)$:

$$\Gamma(A) = \bigoplus_n A^{\otimes n} \tag{3.8.6}$$

which is called second quantization operator acting on the symmetric Fock space

$$(L^2) \cong \bigoplus_{n=0}^{\infty} L^2(\widehat{R^n, \sqrt{n!}d^n u}). \tag{3.8.7}$$

The operator

$$\Gamma(A)^p = \Gamma(A^p), \ p \geq 1, \tag{3.8.8}$$

also acts on (L^2). Its domain is dense in (L^2).

We now set

$$(S_p) = D(\Gamma(A^p)) \quad (\subset (L^2)),$$

where D means the domain. The space (S_p) is a Hilbert space with norm $\| \cdot \|_p$. We have

$$\cdots \subset (S_{p+1}) \subset (S_p) \subset \cdots \subset (L^2) \subset \cdots \subset (S_{-p}) \subset (S_{-(p+1)}) \subset \cdots$$

where (S_{-p}) denotes the dual space of (S_p).

We can see that the system $\{\|\cdot\|_k, k \in Z\}$ is consistent and that the injection from (S_{k+1}) to (S_k) is of Hilbert-Schmidt type for every $k \in Z$. Define the space of *test functional* as the projective limit of the $(S_p), p \geq 0$.

$$(S) = \bigcap_p (S_p) \tag{3.8.9}$$

for which the projective limit topology is introduced. It can be seen that (S) is a countably Hilbert and nuclear space. The dual space $(S)^*$ of (S) is therefore given by

$$(S)^* = \bigcup_p (S_p^*). \tag{3.8.10}$$

A member of $(S)^*$ is a *generalized white noise functional.* Actually, this is the second definition of a generalized white noise functional. We also have a Gel'fand triple

$$(S) \subset (L^2) \subset (S)^*.$$

The following assertions can easily be proved.

Proposition 3.10 *(S) is dense in (L^2).*

Proposition 3.11 *(S) is an algebra.*

Definition 3.7 Let $U(\xi)$ be a complex-valued functional on S. U is called a U-functional if the following two conditions are satisfied.

1) For all $\xi, \eta \in S$, the mapping $\lambda \mapsto U(\eta + \lambda\xi)$, $\lambda \in R$ has an entire analytic extension, denoted by $U(\eta + \lambda z), z \in \mathbf{C}$.
2) There exist $p \in N$ and $c_1, c_2 > 0$ such that for all $z \in \mathbf{C}, \xi \in S$, the following inequality holds.

$$|U(z\xi)| \leq c_1 e^{c_2 |z|^2 |\xi|_p^2}.$$

Theorem 3.4 *(Potthoff-Streit[50]) If $\varphi \in (S)^*$, then $S\varphi$ is a U-functional and conversely, for a U-functional U, there exists a unique $\varphi \in (S)^*$, such that $S\varphi = U$.*

For the proof we refer to the monograph[50].

Example 3.8
i) $H_n(\dot{B}(t), \frac{1}{dt}), n \geq 1$ is a member of both $\mathcal{H}_n^{(-n)}$ and $(S)^*$.

ii) $\dot{B}(t)\dot{B}(s), t \neq s$, the multiplication is defined and the product is in $\mathcal{H}_2^{(-2)}$. (see Example 3.3.)

Proposition 3.12 *Exponential function* $e^{a\langle x,\xi\rangle}$, *for any* $a \in \mathbf{C}$ *and* $\xi \in E$ *is a member of both* $(L^2)^+$ *and* (S).

Example 3.9 Let $\varphi(x) \in \mathcal{H}_2$ and consider $f(x) = e^{i\varphi(x)}$, which is in (L^2). Let $F(u,v)$ be the kernel function of the integral representation of $\varphi(x)$ and the eigen system corresponds to the kernel F be $\{\lambda_n, n \geq 1\}$ and $\{\lambda_n, n \geq 1\}$. Then, we have

$$(Sf)(\xi) = C(\xi)\delta(2;F)^{-\frac{1}{2}} \exp\left[\sum_{n=1}^{\infty} \frac{i}{-\lambda_n + 2}(\eta_n, \xi)^2\right].$$

3.9 Approximation to Hida distributions

When we discuss white noise functional $\varphi(\dot{B}(t), t \in R^1)$, it is natural to take its approximation of the form $f(\frac{\Delta_j B}{\Delta_j}, j \in Z)$, where $\Delta_j B$ is the variation of a Brownian motion $B(t)$ over the interval Δ_j, and where $\{\Delta_j, j \in Z\}$ is a partition of R^1.

Proposition 3.13 $\frac{\Delta B}{\Delta} \to \dot{B}(t)$ *as* $\Delta \to \{t\}$ *in the topology of* $\mathcal{H}_1^{(-1)}$.

Proof. We know that $\frac{\Delta B}{\Delta} \in \mathcal{H}_1$.

$$\frac{\Delta B}{\Delta} \longleftrightarrow \frac{1}{\Delta}\chi_\Delta$$

and

$$\dot{B}(t) \longleftrightarrow \delta_t.$$

Thus we have to show that in $K_1^{(-1)}(R^1)$

$$\frac{1}{\Delta}\chi_\Delta \to \delta_t.$$

Using their Fourier transform, we can easily show

$$\int_{-\infty}^{\infty} \frac{\left| \frac{1}{\epsilon}\int_{t-\frac{\epsilon}{2}}^{t+\frac{\epsilon}{2}} e^{i\lambda u} du - e^{i\lambda t} \right|^2}{1+\lambda^2} d\lambda \to 0 \tag{3.9.1}$$

by bounded convergence theorem. Thus, the assertion follows.

■

We have proved that $\frac{\Delta B}{\Delta} \to \dot{B}(t)$ in the topology of $\mathcal{H}_1^{(-1)}$ (not in (L^2)). We then come to observations on the convergence of higher degree polynomials in $\frac{\Delta_k B}{\Delta_k}$, where the Δ_k's are disjoint.

Theorem 3.5 *A monomial* $(\frac{\Delta B}{\Delta})^n, n > 0$, *does not converge in the space* $(L^2)^-$. *But a Hermite polynomial* $H_n(\frac{\Delta B}{\Delta}; \frac{1}{\Delta})$ *converges to* $\frac{1}{n!} : \dot{B}(t)^n :$ *in the space* $(L^2)^-$, *where* $: \dot{B}(t)^n :$ *is a member of* $\mathcal{H}_n^{(-n)}$ *and is defined to be such that its S-transform is* $\xi(t)^n$.

Proof. The increment of a Brownian motion ΔB can be realized as $\langle x, \chi_\Delta \rangle$ in the white noise space (E^*, μ). We take the Hermite polynomial with parameter $\frac{1}{\Delta}$ with $n!$ in front such that $n! H_n(\frac{\Delta B}{\Delta}; \frac{1}{\Delta})$, which is equal to $(2\Delta)^{-n/2} H_n(\frac{\Delta B}{\sqrt{2\Delta}})$. We denote it by $\varphi_\Delta(x), x \in E^*(\mu)$.

The $\varphi_\Delta(x)$ is an ordinary functional, we shall however discuss its limit, as $\Delta \to \{t\}$ in the space of Hida distributions. For this purpose we take a test functional $\exp[i\langle x, \xi \rangle]$ and observe the bilinear form

$$\langle \varphi_\Delta, \exp[i\langle x.\xi \rangle] \rangle.$$

By using the formulas (A.2.19) and (A.2.25), this bilinear form turns out to be the $\mathcal{T}$-transform expressed in the form

$$\int e^{i\langle x.\xi \rangle} (2\Delta)^{-\frac{n}{2}} H_n(\frac{\Delta B}{\sqrt{2\Delta}}) d\mu(x),$$

where H_n is the ordinary Hermite polynomial of degree n.

By the actual computation, the above formula turns up to be

$$C(\xi) i^n \langle \xi, \frac{\chi_\Delta}{\Delta} \rangle^n,$$

which tends to

$$C(\xi) i^n \xi(t)^n.$$

We know that $\xi(t)^n$ is a linear function of the test functional $e^{i\langle x,\xi\rangle}$.

Thus we have proved the theorem. ■

Corollary 3.2 *Consider a product of Hermite polynomials*

$$\prod_k H_{n_k}\left(\frac{\Delta_k B}{\Delta_k} : \frac{1}{\Delta_k}\right)$$

where Δ_k's are non-overlapping and tending to the point sets $\{t_k\}$'s. Then, the product converges to $:\Pi\dot{B}(t_k)^{n_k}:$ in $(L^2)^-$.

There is the second approximation which has been used frequently, but it is not in line with our approach. We explain it briefly as follows.

Take a complete orthonormal system $\{\xi_n\}$ such that each $\xi_n \in E$. Then, $\{X_n = \dot{B}(\xi_n)\}$ may be taken to be the system of variables of white noise functional. Namely, we can discuss functions of the form $f(X_n, n \in Z)$ in the discrete expression, and it approximates desired functional. The disadvantage is that there is no way to define a sequence to converge to $\dot{B}(t)$. In addition, it is impossible to express the phenomena that are developing, since t is not used explicitly.

What have been discussed are concerned with the passage from discrete to continuous, from which we have to proceed to the passage of operators. Concerning the limit of operators, we have to provide lots of preparations and the method is complicated. We shall, therefore, discuss in the separate section.

Note that the method of the approximation implies the definition of some generalized functionals and implicitly suggests the idea of renormalization which will be discussed in the next section.

3.10 Renormalization in Hida distribution theory

The renormalization is one of the most important techniques in white noise analysis; indeed, it is a key tool for introducing generalized white noise functionals, the so-called Hida distributions. One may think that it comes from the similar idea to the "renormalization" of quantum dynamics in physics, but in reality it is different, although we use the same term as in physics.

The first and in fact main subject to be presented in this section is a general setting of renormalization. Then, naturally follow various interpretations and applications.

We remind that the system of variables $\{\dot{B}(t), t \in R^1\}$ is taken to be the variable system of random functionals and is fixed. As in the ordinary elementary calculus, we have to determine the class of functions to be analyzed. It is quite reasonable, first of all, to choose polynomials in the given variables. Unfortunately, they are not ordinary white noise functionals if the degree is greater than 1, they are not even generalized functionals without modifications.

On the other hand, the spaces $(L^2)^-$ and $(S)^*$ of generalized white noise functionals have been established and fixed.

These two facts should not be contradicting. Under this situation, we are going to use the technique "renormalization" to overcome the difficulty in a reasonable manner.

Ideas

Variables $\{\dot{B}(t), t \in R^1\}$ have been fixed. It is therefore natural, as was mentioned before, to start with polynomials in the $\dot{B}(t)$'s and their exponential. However, it is known that polynomials are not always ordinary (generalized) white noise functionals, although they are the most basic functions. We are, therefore, requested to have them modified to invite them to manageable class, namely to $(L^2)^-$ or to $(S)^*$. This can be done by, so-to-speak, *renormalization* which is going to be explained in what follows.

We start with the linear space $\mathcal{A}$ spanned by all the polynomials in $\dot{B}(t), t \in R^1$ over the complex number field $\mathbf{C}$.

Proposition 3.14 *The set $\mathcal{A}$ forms a graded algebra:*

$$\mathcal{A} = \sum_{n=0}^{\infty} \mathcal{A}_n,$$

where $\mathcal{A}_n$ is spanned by the polynomials of degree n.

The grade is obviously the degree of a polynomial. We can therefore define the annihilation operator ∂_t that makes the degree of a polynomial in $\dot{B}(t)$ decrease by one. As the dual operator (in a sense) we can define the creation operator ∂_t^* that makes the degree of a polynomial in $\dot{B}(t)$ increase by one.

The above operators should be discussed separately, but here it is to be mentioned that the algebra is graded, and ∂_t, ∂_t^* are concerned with grade.

Thus, we are given a non-commutative algebra generated by annihilation and creation operators in line with the study of operator algebra.

A general theory for renormalizations

To establish a general theory of renormalization, we have to note the characteristic properties of $\dot{B}(t)$'s.

1) Each $\dot{B}(t)$ is elemental (atomic) ; hence, it is natural to remind the idea of **reduction** of random functions.
2) It is, intuitively speaking, $\frac{1}{\sqrt{dt}}$ in size. Thus, white noise analysis is quite different from non-random calculus and even from discrete calculus. Formally, $\dot{B}(t)$ may be viewed as a stochastic square root of δ_t. Compare Mikusinski's idea of the square of the delta function (see the literature[96]).
3) If the $\dot{B}(t)$ is understood to be a multiplication variable, it has to be the sum of creation and annihilation:
$$\dot{B}(t) = \partial_t + \partial_t^*,$$
so that we immediately see that its powers generate an algebra involving non-commutative operators, and so on.

With these facts in mind, we shall propose a method of renormalization starting from the algebra $\mathcal{A}$ as general as possible.

The key role is played by the so-called $\mathcal{S}$-transform in white noise theory. In fact, the S-transform of a white noise functional $\varphi(x)$ defined by (3.5.2) is

$$(\mathcal{S}\varphi)(\xi) = e^{-\frac{1}{2}\|\xi\|^2}\langle e^{\langle x,\xi\rangle}, \varphi(x)\rangle. \qquad (3.10.1)$$

The reasons why we use the $\mathcal{S}$-transform are now in order.

a) In order to have generalized functionals, we have to take a test functional space. The function $\exp[\langle x, \xi\rangle]$ is not only a test functional by itself (see Proposition 3.12), but also a generator of ordinary white noise functionals.

b) The renormalization, that we are going to introduce, could be, formally speaking, a projection of $\mathcal{A}$ down to the space $(L^2)^-$. The inner product of the exponential functional and a polynomial defines a *projection*, since the exponentials generate the space of test functionals.

c) The method to be proposed should be applied not only to $\mathcal{A}$, but also to exponential functionals by the same principle.

d) Any member in $\mathcal{A}$ should be modified so that the modified function is a Hida distribution for which the inner product with $e^{\langle x,\xi\rangle}$ (i.e. the S-transform of which) is necessary continuous.

With these considerations, we define the *additive renormalization* by the following facts.

Proposition 3.15 *i) Let $\varphi(\dot{B}(t))$ be a polynomial in $\dot{B}(t)$. Then, we have*

$$(\mathcal{S}\varphi)(\xi) = p(\xi(t)) + O(\frac{1}{dt}),$$

where p is a polynomial.

ii) For a product $\prod \varphi_j(\dot{B}(t_j))$ of polynomials (using the same notations as in i)), we have the S-transform

$$\prod p_j(\xi(t_j)) + O(\prod \frac{1}{dt_j}).$$

The proof is easy. One thing to be noted is how to understand the symbol $\frac{1}{dt}$. A polynomial in $\dot{B}(t)$ is approximated by that of $\frac{\Delta B}{\Delta}$. Apply the S-transform to find terms involving $\frac{1}{\Delta}$. For computation one may compare the Hermite polynomials with parameter. (σ^2 is replaced by $\frac{1}{dt}$ or $\frac{1}{\Delta}$.)

Some more interpretation is as follows. We wish to clarify the meaning of renormalization in white noise analysis. It is necessary to determine the exact quantity of the term $O(\prod \frac{1}{dt})$ which should be subtracted.

First, we think of renormalization for $\dot{B}(t)^p$. That is, $\dot{B}(t)^p$ is to be modified by subtracting off "infinity" which is to be understood as a polynomial in $\frac{1}{dt}$. The coefficients in the polynomial are powers of $\dot{B}(t)$. Then the modified polynomial is a Hida distribution that is a continuous linear functional of test functionals, in particular, $e^{i\langle \dot{B},\xi\rangle}$. On the other hand, we know that

$$E[e^{i\langle \dot{B},\xi\rangle}(\frac{\Delta B}{\Delta})^p]$$

is expressed in the polynomial of the form

$$\langle \xi, \chi_\Delta\rangle^p + a_1(\xi)\frac{1}{\Delta} + \cdots + a_p(\xi)(\frac{1}{\Delta})^p.$$

Then, the exact quantity to be subtracted off, namely

$$\sum_{k=1}^{p} a_k(\xi)(\frac{1}{\Delta})^k.$$

Practically, first subtract $a_p(\xi)(\frac{1}{\Delta})^p$, then $a_{p-1}(\xi)(\frac{1}{\Delta})^{p-1}$ and so on in an order of magnitude.

Thus, this renormalization asserts not simply a modification, but also determines the exact amount to be subtracted.

For a general polynomial of $\dot{B}(t)$'s, the renormalization is to be done with the same idea as above.

We define the renormalization using the notation $: \cdot :$ by

$$: \prod \varphi_j(\dot{B}(t_j)) := \mathcal{S}^{-1}(\prod p_j(\xi(t_j))). \tag{3.10.2}$$

Theorem 3.6 *The operation $: \cdot :$ can be extended linearly to $\mathcal{A}$ and satisfies*

i) it is idempotent.

ii) it can be extended to exponential functionals of quadratic forms of the $\dot{B}(t)$'s.

Note. It is difficult to say that the operation $: \cdot :$ is Hermitian, but $: \cdot :$ satisfies partly a role of the projection operator.

Example 3.10 By using (A.2.26), we have

$$: \dot{B}(t)^n := n! H_n(\dot{B}, \frac{1}{dt}).$$

Exponentials of quadratic functionals.

Renormalization can be applied to those exponential functionals through the power series expansion. According to the well-known Potthoff-Streit characterization of $(S)^*$-functionals, we essentially need to think of the case where the exponent is a polynomial in $\dot{B}(t)$'s of degree at most two. Linear exponent is easily dealt with, so that we shall be involved only in exponential functionals with quadratic exponent.

Recall that the result on the S-transforms of exponential functions of "ordinary" quadratic functionals of white noise.

Take $x \in E^*(\mu)$ and let $\varphi(x)$ be a real-valued $\mathcal{H}_2$-functional with kernel F. Set

$$f(x) = \exp[\varphi(x)].$$

Then, we have the following theorem.

Theorem 3.7 *Suppose that the kernel F has no eigenvalues in the interval $(0, 4]$. Then*

$$(\mathcal{S}f)(\xi) = \delta(2i; F)^{-\frac{1}{2}} \exp\left[\int\int \hat{F}(u,v)\xi(u)\xi(v)dudv\right], \qquad (3.10.3)$$

where $\hat{F} = \frac{F}{-I+2F}$, ($F$ being an integral operator), and where $\delta(2; F)$ is the modified Fredholm determinant.

Proof. See (3.10.1) and the book[30] Section 4.6, in particular. Also see Smithies[126]. ∎

Note. One may ask why the modified Fredholm determinant is used instead of the Fredholm determinant. The answer is worth to be mentioned. Roughly speaking, $f(x)$ is not quite equal to a quadratic form, but it is a renormalized quadratic form. The diagonal term is subtracted off from the ordinary expression of quadratic form. This fact is related to the modification of the determinants that appear in the expansion of the Fredholm determinant.

Now we have the theorem

Theorem 3.8 *The renormalization of a functional of the form $f(x) = \exp[\varphi(x)]$ with quadratic functional φ is necessary only when φ needs to be* *renormalized. In such a case we have the renormalization* $: f(x) :$ *of* $f(x)$ *just by deleting the factor* $\delta(2i; F)^{-\frac{1}{2}}$ *of the formula in Theorem 3.7.*

For the proof it is recommended again to use the formulas for Hermite polynomials with parameter in the Appendix.

We can now come to a generalized functional. To be precise, let the kernel F of $\varphi(x)$ be a member of the Sobolev space $K^{-3/2}(R^2)$, say F^-. We can find a sequence $\{F_n\}$ in $L^2(R^2)$ such that

$$F_n \to F^-$$

in the $K^{-3/2}(R^2)$-topology. Hence

$$\int\int F_n(u,v)\xi(u)\xi(v)dudv \to \int\int F^-(u,v)\xi(u)\xi(v)dudv.$$

(The right-hand side is better to be written as $\langle F^-, \xi \otimes \xi \rangle$, since F^- is a generalized function.)

On the other hand, the constant in front, the modified Fredholm determinant denoted by $\delta(2; F_n)$, tends to infinity. This amount should be removed. Namely, this is the multiplicative renormalization.

As is seen by the note stated in [2] §4.6, the renormalization defined above is consistent with that of polynomials.

Our aim is to define generalized functionals of the variables $\dot{B}(t)$'s in linear case and quadratic case; we have seen how to have general and rigorous definition of generalized functionals of white noise $\dot{B}(t)$'s.

For this aim, similar to the definition of ordinary (non-random) generalized functionals, we have to determine a class of generalized functionals.

Our situation is

1) $\{\dot{B}(t), t \in R\}$ is total in $\mathcal{H}_1^{-1}$. Each $\dot{B}(t)$ is well defined and is a member of $\mathcal{H}_1^{-1}$.
2) In addition, space of test functionals should involve $\{e^{i\langle \dot{B}, \xi\rangle}, \xi \in E\}$ the span of which is dense in (L^2). We define a Gel'fand triple for which the basic space is (L^2); namely

$$(L^2)^+ \subset (L^2) \subset (L^2)^-.$$

3) $(L^2)^-$ should be defined so that all the renormalized polynomials in $\dot{B}(t)$'s are included. The general definition of renormalization is to be defined by the choice of test functionals. In other words $(L^2)^+$ and $(L^2)^-$ are determined in the following manner.

i) Required space of distributions of the $\dot{B}(t)$'s should be spanned by the renormalized Hermite polynomials of the form

$$\prod_k H_{n_k}(\dot{B}(t_k); \frac{1}{dt}).$$

So we first think of the bilinear form

$$\langle e^{i\langle \dot{B}, \xi\rangle}, H_{n_k}(\dot{B}(t_k); \frac{1}{dt})\rangle \tag{3.10.4}$$

which should be continuous in any test functionals $e^{i\langle \dot{B}, \xi\rangle}, \xi \in E$. The bilinear form (3.10.4) is just the $\mathcal{T}$ transform of the Hermite polynomial. The result is $\xi(t)^n$. This should be a continuous function of $e^{i\langle \dot{B}, \xi\rangle}$.

Note that (3.10.4) can be defined by approximation of Hermite polynomial.

ii) $\xi(t)^n$ is continuous in Sobolev norm $K^1(R^n)$, since $\xi(t)$ is continuous in Sobolev norm $K^1(R^1)$, for

$$\begin{aligned}
\xi(t) &= \langle \xi, \delta_t\rangle = \langle \hat{\xi}, \hat{\delta}_t\rangle \\
&= \frac{1}{\sqrt{2\pi}} \int \hat{\xi}(\lambda) e^{-it\lambda} d\lambda| \\
|\xi(t)| &= \frac{1}{\sqrt{2\pi}} |\int \hat{\xi}(\lambda) e^{-it\lambda} d\lambda| \\
&= \frac{1}{\sqrt{2\pi}} |\int \sqrt{1+\lambda^2}\hat{\xi}(\lambda) \frac{e^{-it\lambda}}{\sqrt{1+\lambda^2}} d\lambda| \\
&= \frac{1}{\sqrt{2\pi}} \sqrt{\int (1+\lambda^2)|\hat{\xi}(\lambda)|^2 \int \frac{1}{1+\lambda^2} d\lambda} \\
&\le \text{const} \cdot \|\xi\|_1^2.
\end{aligned}$$

If we restrict the branch by $e^0 = 1$, the $\langle \dot{B}, \xi\rangle$ is continuous in $e^{i\langle \dot{B}, \xi\rangle}$.

iii) We have proved that the $H_n(\dot{B}, \frac{1}{dt})$ is the renormalized function of $\dot{B}(t)^n$, thus the Hermite polynomial with parameter $\frac{1}{dt}$ is in $(L^2)^-$.

iv) Renormalization of $\prod_{t_j:\text{different}}(\dot{B}(t_j)^{n_j}$ is done multiplicatively to obtain $\prod_{tj:\text{ different}} H_{n_j}(\dot{B}(t_j), \frac{1}{dt_k})$.

v) The system of Hermite polynomials $\{\prod_{tj:\text{different}} H_{n_j}(\dot{B}(t_j), \frac{1}{dt_j})\}$ is total in $(L^2)^-$.

vi) As for the exponential functionals of $\dot{B}(t)$'s, we only need to consider functionals of $\exp[\varphi(\dot{B})]$, when $\varphi(\dot{B})$ is a polynomial of degree at most 2. So, we apply the known results.

We have noticed in Section 3.4 for the renormalization of quadratic forms, in particular normal functionals.

Now we have come to the position to remind the exponential functionals of $\dot{B}(t)$'s, e.g. an application to the Feynman integrals, the Gauss kernels and others.

Multiplicative renormalization will be illustrated by the following two examples. The first one is an exponential of linear functionals of $\dot{B}(t)$'s.

A functional $\exp[\int \xi(u)\dot{B}(u)du]$ is an ordinary functional, so that no modification is to be done. However, for the exponential of $\mathcal{H}_1$-functional, we may need to modify. Example 3.12 explains this fact. While Example 3.11 requires more extensive renormalization.

Example 3.11 Set

$$\varphi(\dot{B}(t)) = e^{\dot{B}(t)}$$

which is a formal expression.

Apply the formal power series expression

$$\sum \frac{1}{n!}\dot{B}(t)^n.$$

Apply term by term renormalization (i.e. in fact additive renormalization), then we have

$$\sum \frac{1}{n!} : \dot{B}(t)^n :$$

which gives us

$$: \varphi(\dot{B}) := e^{\dot{B}(t) - \frac{1}{2dt}}.$$

So that its S-transform is

$$S(:\varphi(\dot{B}):)(\xi) = \sum \frac{1}{n!}\xi(t)^n = e^{\xi(t)},$$

which exists in the space $\boldsymbol{F} = S((L^2))$.

Example 3.12 The Gauss kernel

We begin with a formal expression $\exp[c \int \dot{B}(t)^2 dt]$, which is not rigorously defined. But after the multiplicative renormalization we have

$$S(:\exp[c\int \dot{B}(t)^2 dt]:)(\xi) = \exp[\frac{c}{1-2c}\|\xi\|^2],$$

where $c = \frac{1}{2}$ is excluded.

Approximate the integral by $\sum(\frac{\Delta B}{\Delta_n})^2$ and apply the S-transform, to obtain a formula

$$C(\{\Delta_k^n\})e^{\frac{c}{1-2c}}\sum_k(\Delta_k^n\xi)^2\Delta_k$$

where $C(\{\Delta_k\})$ tends to infinity, as the partition $\{\Delta_k^n\}$ is getting finer.

Thus, the limit $C(\{\Delta_k^n\})$, which is infinity, is the amount to be decided as the number of multiplicative renormalization. Then

$$S(:e^{c\int \dot{B}(t)^2 dt}:)(\xi) = e^{\frac{c}{1-2c}\int \xi(u)^2 du}.$$

From the probabilistic side, we claim that renormalization, either additive or multiplicative, occurs necessarily in the case where we deal with continuously many independent idealized random variables. As far as countably many independent members are concerned, we do not need any renormalization.

For the Wick product of $\prod_j x(u_j)$ in the terminology of physics, the formulas are as follows. They can be compared with our renormalizations where $x(u_j) = \dot{B}(u_j)$.

$$:x(u_1): = x(u_1) \tag{3.10.5}$$

$$:x(u_1)x(u_2): = x(u_1)x(u_2) - \delta(u_2 - u_1)x(u_1) - x(u_1) \tag{3.10.6}$$

$$:\prod_{j=1}^{n} x(u_j): = x(u_n):\prod_{j=1}^{n-1} x(u_j): - \sum_{k=1}^{n-1}\delta(u_n - u_k):\prod_{j\neq k}^{n-1} x(u_j): \tag{3.10.7}$$

We continue to state some notes on the renormalization.

Example 3.13 First, we take an exponential functional of a linear function $l(\dot{B})$ of the form $\exp[l(\dot{B})]$.

We have formal power series

$$\exp[l(\dot{B})] = \sum \varphi_n(\dot{B}).$$

We apply the additive renormalization to each term, If the sum of the renormalized functionals converges in $(L^2)^-$, then the sum is the renormalized exponential functionals.

Remark 3.4 *By the Potthoff-Streit characterization theorem of generalized white noise functionals, we do not need to consider exponential of higher degree polynomials.*

Example 3.14 For $\dot{B}(t)^2\dot{B}(s), t \neq s$ the term to be subtracted is $\frac{1}{dt}\dot{B}(s)$ to have $: \dot{B}(t)^2\dot{B}(s) :$.

Chapter 4

White Noise Analysis

The main purpose of this chapter is to establish the analysis of white noise functionals, specifically we are interested in the calculus of generalized white noise functionals. The calculus is often called Hida calculus.

We shall first define the partial differential operators in the variables $\dot{B}(t)$'s acting on the functionals $\varphi(\dot{B}(t), t \in R^1)$. Note that we have already briefly introduced the calculus in the discrete parameter case in Chapter 1. Thus, it is better to start with the review of the method that the passage to the continuous parameter case from discrete parameter case where the variables are at most countable. We shall see not only refinement of discrete parameter case, but also necessary essential development in tools and idea to discuss functionals of $\dot{B}(t)$.

As soon as we grade up to the continuous parameter case, taking the set $\{\dot{B}(t)\}$ to be the variable system, one can see the significant characteristics of white noise analysis where Hida distributions are involved.

Another particular characteristic of white noise analysis is that it has an aspect of the *harmonic analysis*, arising from the infinite dimensional rotation group which will be briefly mentioned in Section 4.4. The white noise measure μ is invariant under the infinite dimensional rotations. Furthermore, the rotation group characterizes the white noise measure μ. This fact is certainly a big advantage that makes the white noise analysis profound.

Since each variable is atomic and since the collection of variables is parametrized by t that runs through R^1, the time development can be represented explicitly in terms of $\dot{B}(t)$'s without smearing them. We shall naturally be led to the case of random phenomena which are parametrized by field variables. We shall therefore move to the case of random fields. Although the analysis becomes complex, we can find interesting properties of random fields, which depend on the geometric structure of multi-dimensional parameter space.

4.1 Operators acting on $(L^2)^-$

Various tools will be employed for the analysis of white noise functionals. Basic operators are partial differential operators in the variable $\dot{B}(t)$'s and their adjoints. Then follow Laplacians and related operators.

Since $\dot{B}(t)$ is a variable of white noise functional, it is natural to define the partial derivative in $\dot{B}(t)$.

As for the operand, or the domain of the partial differential operators, we take the space $(L^2)^-$ of generalized functionals of the $\dot{B}(t)$'s. The main reason is that the polynomials in $\dot{B}(t)$'s span the entire space $(L^2)^-$ which is a Hilbert space.

For a rigorous definition of partial differential operators $\frac{\partial}{\partial \dot{B}(t)}$, we use the representation of white noise functionals by using the S-transform. We therefore have to review the theory of functional analysis to analyze the U-functionals.

Let us apply the S-transform to a white noise functional $\varphi \in (L^2)^-$ to have U-functionals which is denoted by $U(\xi)$. Recall that

$$(S\varphi)(\xi) = \exp[-\frac{1}{2}\|\xi\|^2] \int e^{i\langle x,\xi\rangle}\varphi(x)d\mu(x).$$

Let E be a nuclear space. The variation of $U(\xi)$, denoted by $\delta U(\xi)$ is defined as follows, where $\delta\xi$ is infinitesimal in the sense that $\|\delta\xi\|_n$ is small for some norm $\|\cdot\|_n$, among $\{\|\cdot\|_k, k \in Z\}$ defining the topology of E.

Definition 4.1 If the variation $\Delta U(\xi)$ is expressed in the form

$$\Delta U(\xi) = U(\xi + \delta\xi) - U(\xi) = \delta U(\xi, \delta\xi) + o(\delta\xi),$$

where $\delta U(\xi, \delta\xi)$ is linear in $\delta\xi$ and where $o(\delta\xi)$ denotes the term of smaller order than $\delta\xi$ in E-topology, then $\delta U(\xi, \delta\xi)$ is *the first variation* of $U(\xi)$. If, further, there exists a generalized function $U'(\xi, t) = U'(\xi)$ of t in E^*, such that

$$\delta U(\xi, \delta\xi) = \langle U'(\xi), \delta\xi\rangle \left(= \int U'_\xi(\xi, t)\delta\xi(t)dt \right), \tag{4.1.1}$$

then $U'(\xi)$ is called the *first functional derivative* in the sense of *Fréchet derivative* of $U(\xi)$. In this case, $U(\xi)$ is said to be Fréchet differentiable.

The Fréchet derivative $U'(\xi)$ is often denoted by

$$\frac{\delta}{\delta\xi(t)}(S\varphi).$$

The expression

$$\int U'_\xi(\xi,t)\delta\xi(t)dt$$

is called the *Volterra form* of the variation of $U(\xi)$.

Here we note that $\delta\xi(t)$ is the differential of $\xi(t)$ and is running through E. It is a continuous analogue of the differential du of $u(x_1, x_2, \cdots, x_n)$:

$$du = \sum_1^n \frac{\partial u}{\partial x_j} dx_j.$$

For the idea of defining the functional derivative, we refer to Volterra[131] Chap. II and P. L'evy [1951] Part I.

Note that $U'_\xi(\xi,t)$ is a functional of ξ and a generalized function of t.

Definition 4.2 If $U'(\xi,t)$ has a functional derivative, which is denoted by $U''(\xi,t,s)$, then it is the *second functional derivative* of U and the functional $U(\xi)$ itself is said to be second Fréchet differentiable in the topology of E^*.

We tacitly assume that $U''(\xi,t,s)$ is measurable in (t,s).

If the second functional derivative of $U(\xi)$ exists then the second variation $\delta^2 U(\xi,\delta\xi)$ is expressed in the form

$$\delta^2 U(\xi,\delta\xi) = \int\int F(\xi,t,s)\delta\xi(t)\delta\xi(s), \qquad (4.1.2)$$

where $F(\xi,t,s)$ is assumed to be in the symmetric Sobolev space $K^{\widehat{(-\frac{3}{2})}(R^2)}$ for every ξ. Note that $F(\xi,t,s)$ may be a generalized function of (t,s).

It is noted that $\delta^2 U(\xi,\delta\xi)$ is viewed as a quadratic form of $\delta\xi$, where ξ is fixed.

There is a favorable case (P. Lévy[85]), we may restrict our attention to the functional $U(\xi,t)$ such that the second variation is expressed in the following. There F in the expression (4.1.2) may be decomposed into two parts, to have

$$\delta^2 U(\xi, \delta\xi) = \int\int F(\xi, t, s)\delta\xi(t)\delta\xi(s) + \int f(\xi, t)(\delta\xi(t))^2 dt, \qquad (4.1.3)$$

where we assume

$$F(\xi, t, s) \in \widehat{L^2(R^2)}$$

and

$$f(\xi, t) = f\left(\xi, \frac{t+s}{2}\right)\delta(t-s), \quad f \in L^2(R^1),$$

for every $\xi \in E$.

Namely, we are interested in the class of the second derivatives that can be expressed in the form (4.1.3). The functional $U(\xi)$ is called twice differentiable functional in Fréchet sense.

Following P. Lévy, a quadratic functional of $\delta\xi$ expressed in the form (4.1.3) is called a normal functional, the first term and the second term of which are called the regular term and the singular term, respectively. A normal functional can be decomposed uniquely as in (4.1.3).

Before defining the partial differential operators, we have to ask what does derivative mean when variables of functions are given. One of the reasonable answer is as follows. Since the given variables are $\dot{B}(t)$'s, polynomials of which are the basic functions. Differential operator should act as an annihilation operator in such a way that the degree of the polynomial decreases by one. We might expect $\dot{B}(t)^n \longrightarrow n\dot{B}(t)^{n-1}$ as in the discrete parameter case, i.e. $X^n \longrightarrow nX^{n-1}$. There arises a difficulty that $\dot{B}(t)^n$ is not within our hand. We have renormalized $\dot{B}(t)^n$ to be $:\dot{B}(t)^n:$, the S-transform of which is $\xi(t)^n$. We know that the functional derivative makes the degree to be $n-1$. More precisely, a linear functional of the form

$$U(\xi) = \int f(u)\xi(u)^n du$$

has the functional derivative

$$\frac{\delta}{\delta\xi(t)} U(\xi) = nf(t)\xi(t)^{n-1},$$

which is associated with $U(\xi)$. This means that

$$:\dot{B}(t)^n: \longrightarrow n:\dot{B}(t)^{n-1}: \frac{1}{dt},$$

where $\frac{1}{dt}$ is understood to be $\delta_t(t)$. The mapping can be denoted by $\frac{\partial}{\partial \dot{B}(t)}$.

Further we claim that for a general Hermite polynomials

$$\varphi(\dot{B}) = \prod_j :\dot{B}(t)^{n_j}:,$$

we can show

$$\partial_{t_j}\varphi(\dot{B}) = (n_j - 1):\dot{B}(t_j)^{n_j} \frac{1}{dt_j} \prod_{k\neq j} :\dot{B}(t_k)^{n_k}:.$$

In terms of the S-transform we have

$$\frac{\delta}{\delta\xi(t_j)}U(\xi) = S(\partial_{t_j}\varphi)(\xi),$$

where $U(\xi) = (S\varphi)(\xi)$.

We know that the collection of general Hermite polynomials in $\dot{B}(t)$'s is total in the space $(L^2)^-$.

Thus, the partial derivative is defined as follows.

Definition 4.3 Let $(S\varphi)(\xi) = U(\xi)$ for $\varphi \in (L^2)^-$. If $U'(\xi, u)$ is the S-transform of a generalized white noise functional $\varphi'(x, u)$ for almost all u, then φ is said to be $\dot{B}(t)$-*differentiable* and we define ∂_t by

$$\partial_t\varphi(x) = \varphi'(x, t)$$

and ∂_t is called the partial differential operator. The φ' is the partial derivative of φ.

The operator ∂_t is understood to be

$$\partial_t = \frac{\partial}{\partial \dot{B}(t)}, \tag{4.1.4}$$

which is often called *Hida derivative.*

Remark 4.1 *In the above definition we take Fréchet derivative. The Gâteau derivative may fit for the discrete case, which is different from what we are concerned with.*

The Hida derivative given by Definition 4.3 as well as (4.1.4) can be rewritten as

$$\partial_t \varphi(x) = S^{-1}\left(\frac{\delta}{\delta\xi(t)}(S\varphi)(\xi)\right)(x). \tag{4.1.5}$$

To make this definition acceptable we have to add the following two conditions.

i) The functional derivative $\frac{\delta}{\delta\xi(t)}$ is the Fréchet derivative.
ii) $(S\varphi)(\xi)$ is in the domain of $\frac{\delta}{\delta\xi(t)}$ and $\frac{\delta}{\delta\xi(t)}U(\xi)$ is the S-transform of some φ'_t in $(L^2)^-$.

If all the requirements are satisfied, we say that φ is $\dot{B}(t)$-differentiable or φ is in the domain $\mathcal{D}(\partial_t)$ of ∂_t.

The following assertion follows straightforwardly from the definition.

Proposition 4.1 *The domain $\mathcal{D}(\partial_t)$ is dense in $(L^2)^-$.*

The readers can realize this assertion by the examples in the following.

It is noted that this definition of the partial derivative is relevant for white noise calculus, since $\dot{B}(t)$'s are taken to be the variables of functionals in question. Some other reasons will be clarified as the calculus develops. In addition, the derivative is conjugate to the Fréchet derivative through the S-transform.

Example 4.1 If $\varphi = \int_{R^2} F(u,v) : \dot{B}(u)\dot{B}(v) : dudv$ in $\mathcal{H}_2^{(-2)}$, then its S-transform, i.e. the U-functional associated to φ is of the form

$$U(\xi) = \int_{R^2} F(u,v)\xi(u)\xi(v)dudv$$

with smooth symmetric kernel $F(u,v)$. Then φ is $\dot{B}(t)$-differentiable as is shown below.

The variation of U is

$$\delta U = 2\int F(t,u)\xi(u)\delta\xi(u)du,$$

thus

$$U'(\xi) = 2\int F(t,u)\xi(u)du.$$

That is,

$$S(\partial_t \varphi)(\xi) = 2\int F(t,u)\xi(u)du$$

which implies

$$\partial_t \varphi = 2\int F(t,u)\dot{B}(u)du.$$

Example 4.2 If $\varphi =: \dot{B}(t)^n :$, then the U-functional associated to φ is

$$U(\xi) = \xi(t)^n = \int \delta_t(u)\xi(u)^n du,$$

then

$$\delta U(\xi) = \int n\delta_t(u)\xi(u)^{n-1}\delta\xi(u)du.$$

Thus, we have

$$U'(\xi) = n\delta_t(t)\xi(t)^{n-1}.$$

Therefore, if formal expression is used, we have

$$\partial_t \varphi = n : \dot{B}(t)^{n-1} : \frac{1}{dt}. \tag{4.1.6}$$

(See 2 in Section 4.1.)

The above expression (4.1.6) is formal, so that $: \dot{B}(t)^n :, n \geq 1$, is not in $D(\partial_t)$, however the formula involving $\frac{1}{dt}$ is convenient when we meet complex formula in terms of $\dot{B}(t)$'s.

Example 4.3 If $\varphi = \int f(t) : \dot{B}(u)^2 : du$ then the U-functional associated to φ is of the form

$$U(\xi) = \int_R f(t)\xi(t)^2 dt$$

with smooth function $f(t)$. Then φ is $\dot{B}(t)$-differentiable since

$$\delta U(\xi) = \int_R 2f(t)\xi(t)\delta\xi(t)dt$$

exists. Then, we have

$$S(\partial_t\varphi)(\xi) = U'(\xi) = 2f(t)\xi(t)$$

which is the S-transform of $2f(t)\dot{B}(t)$. Namely,

$$\partial_t\varphi = 2f(t)\dot{B}(t).$$

Example 4.4 If $\varphi =: e^{c\int \dot{B}(u)^2 du}:$ then the U-functional associated to φ is of the form

$$U(\xi) = e^{\frac{c}{1-2c}\int \xi(t)^2 dt},$$

then

$$\delta U(\xi) = \int \frac{2c}{1-2c}\xi(t)U(\xi)\delta\xi(t)dt$$

thus

$$S(\partial_t\varphi)(\xi) = U'(\xi) = \frac{2c}{1-2c}\xi(t)U(\xi).$$

Then,

$$\partial_t\varphi = \frac{2c}{1-2c}S^{-1}(\xi(t)U(\xi)).$$

$\partial_t\varphi$ belongs to the space $(L^2)^-$, so that φ belongs to the domain of ∂_t.

Proposition 4.2

i) If φ is in the subspace $\mathcal{H}_n^{(n)}$, then it is $\dot{B}(t)$-differentiable and $\partial_t\varphi$ is in $\mathcal{H}_{n-1}^{(n-1)}$. In fact

$$\partial_t : \mathcal{H}_n^{(n)} \rightarrow \mathcal{H}_{n-1}^{(n-1)}$$

is a surjection.

ii) The domain of ∂_t includes $\mathcal{H}_n^{(n)}, n \geq 0$, and is dense in (L^2).

iii) ∂_t is continuous in $\mathcal{H}_n^{(n)}$

Proof. For the proof of i) we note the following fact.

Let φ be in $\mathcal{H}_n^{(n)}$ and let $F(u_1, u_2, \cdots, u_n) \in K^{\frac{n+1}{2}}$ be the kernel function of φ. That is

$$(S\varphi)(\xi_1, \cdots, \xi_2) = \int_{R^n} F(u_1, \cdots, u_n)\xi(u_1)\cdots\xi(u_n)du^n.$$

Thus,

$$\frac{\delta}{\delta\xi(t)}(S\varphi)(\xi) = n\int_{R^{n-1}} F(u_1, \cdots, u_{n-1}, t)\xi(u_1)\cdots\xi(u_{n-1})du^{n-1}. \quad (4.1.7)$$

$$\begin{aligned}
&\|F(u_1, \cdots, u_{n-1})\|^2 = \|\widehat{F}(\lambda_1, \cdots, \lambda_{n-1})\|^2 \\
&= \int\cdots\int_{R^{n-1}} (1+\lambda_1^2+\cdots+\lambda_{n-1}^2)^{\frac{n}{2}} \left|\int \widehat{F}(\lambda_1, \cdots, \lambda_{n-1}, \lambda_n)e^{it\lambda_n}d\lambda_n\right|^2 \\
&\qquad d\lambda_1\cdots d\lambda_{n-1} \\
&= \int\cdots\int_{R^n} (1+\lambda_1^2+\cdots+\lambda_{n-1}^2)^{\frac{n}{2}} \left|\widehat{F}(\lambda_1, \cdots \lambda_{n-1}, \lambda_n)\right|^2 d\lambda_1\cdots d\lambda_n \\
&\le \int\cdots\int_{R^n} (1+\lambda_1^2+\cdots+\lambda_{n-1}^2+\lambda_n^2)^{\frac{n+1}{2}} \left|\widehat{F}(\lambda_1, \cdots, \lambda_{n-1}, \lambda_n)\right|^2 d\lambda_1\cdots d\lambda_n \\
&= \|F(u_1, \cdots, u_n)\|_n^2,
\end{aligned}$$

Hence the assertion i) follows.

ii) can be shown from i).

iii) is obvious. It comes from the computation for the proof of i). ■

As is seen in the above theorem the differential operator is viewed as an **annihilation** operator.

Theorem 4.1 *The operator ∂_t is a derivation.*

Proof. First note that the algebra generated by exponential functions are dense in (L^2). It is easy to see that exponential functions are differentiable. Thus, it suffices to show that the assertion holds for exponential functions.

Let $\varphi_i, i = 1, 2$, be given by

$$\varphi_i(x) = \exp[\langle x, \eta_i\rangle - \frac{1}{2}\|\eta_i\|^2], i = 1, 2.$$

$$\varphi_1(x)\varphi_2(x) = \exp[\langle x, \eta_1 + \eta_2\rangle - \frac{1}{2}\{\|\eta_1\|^2 + \|\eta_2\|^2\}].$$

Taking S-transform,

$$S(\varphi_1\varphi_2)(\xi) = \exp[\langle \xi, \eta_1 + \eta_2 \rangle],$$

and then take the derivative

$$\frac{\partial}{\partial \xi} S(\varphi_1\varphi_2)(\xi) = (\eta_1 + \eta_2) S(\varphi_1\varphi_2)(\xi).$$

Taking back S^{-1}-transform, we have

$$S^{-1} \frac{\partial}{\partial \xi} S(\varphi_1\varphi_2)(\xi) = (\eta_1 + \eta_2)(\varphi_1\varphi_2)(x).$$

Thus, we have

$$\partial_t(\varphi_1(x) \cdot \varphi_2(x)) = (\eta_1(t) + \eta_2(t))(\varphi_1(x)\varphi_2(x)).$$

This proves that

$$\partial_t(\varphi_1 \cdot \varphi_2) = (\partial_t \varphi_1)\varphi_2 + \varphi_1(\partial_t \varphi_2).$$

This relation can be extended to any pair of general functionals in $D(\partial_t)$, since ∂_t is linear.

■

Notation : The adjoint operator of ∂_t is denoted by ∂_t^*.

Proposition 4.3 *For any φ in $(L^2)^+$ (or in (S)) and for any ψ in $(L^2)^-$ (or in $(S)^*$) there exists a generalized functional ψ_t^* such that*

$$(\partial_t \varphi, \psi) = (\varphi, \psi_t^*), \tag{4.1.8}$$

where ψ_t^* is uniquely determined by ψ and t.

Proof. The left-hand side is a continuous linear functional of φ by Proposition (4.2 iii). Hence there exists ψ_t^* such that $(\partial_t \varphi, \psi)$ is written as (φ, ψ_t^*).

■

Notation. The functional ψ_t^* will be denoted by $\partial_t^* \psi$ since it is linear in ψ.

Proposition 4.4 *The operator ∂_t^* is a continuous injection :*

$$\partial_t^* : \mathcal{H}_n^{(-n)} \to \mathcal{H}_{n+1}^{(-n-1)}.$$

Proof. If $F(u_1, \cdots, u_n)$ is the kernel function of $\varphi \in \mathcal{H}_n^{(-n)}$, then $F(u_1, \cdots, u_n)$ is in the Sobolev space $K^{-\frac{n+1}{2}}(R^n)$ and the kernel function associated to $\partial_t^* \varphi$ is $F(u_1, \cdots, u_n)\delta_t(u_{n+1})$.

The Fourier transform of $F(u_1, \cdots, u_n) \cdot \delta_t(u_{n+1})$ is $\widehat{F}(\lambda_1, \cdots, \lambda_n) \cdot e^{it\lambda_{n+1}}$ and

$$\begin{aligned}
&\int_{R^{n+1}} \frac{|\hat{F}(\lambda_1, \cdots, \lambda_n) \cdot e^{it\lambda_{n+1}}|^2}{(1 + \lambda_1^2 + \cdots + \lambda_{n+1}^2)^{\frac{n+2}{2}}} d\lambda^{n+1} \\
&\le c_n \int_{R^n} \frac{|\hat{F}(\lambda_1, \cdots, \lambda_n)|^2}{(1 + \lambda_1^2 + \cdots + \lambda_n^2)^{\frac{n+1}{2}}} d\lambda^n \\
&< \infty,
\end{aligned}$$

since $F \in K^{-\frac{n+1}{2}}(R^n)$. That is

$$\|F(u_1, \cdots, u_n)\delta_t(u_{n+1})\|_{-(n+1)} \le \sqrt{c_n\ n!}\ \|F\|_{-n}.$$

Thus, $F(u_1, \cdots, u_n)\delta_t(u_{n+1}) \in K^{-\frac{n+2}{2}}(R^n)$ and the same for the symmetrization of the kernel, and hence the assertion follows.

■

In view of this we give :

Definition 4.4 The operator ∂_t^* is called a *creation operator*.

Proposition 4.5 *The S-transform of $\partial_t^* \varphi$ is expressed in the form $U(\xi)\xi(t)$, where $U(\xi) = (S\varphi)(\xi)$.*

Proof. Let φ be in $\mathcal{H}_n^{(-n)}$ such that

$$\varphi(x) = \int_{R^n} F(u_1, \cdots, u_n)x(u_1) \cdots x(u_n)du^n.$$

Then

$$U(\xi) = (S\varphi)(\xi) = \int F(u_1, \cdots, u_n)\xi(u_1) \cdots \xi(u_n)du^n.$$

Therefore we have

$$U(\xi)\xi(t) = \int_{R^{n+1}} F(u_1, \cdots, u_n)\delta_t(u_{n+1})\xi(u_1) \cdots \xi(u_n)\xi(u_{n+1})du^{n+1}.$$

On the other hand,

$$\begin{aligned}\partial_t^*\varphi &= \partial_t^*\left(\int_{R^n} F(u_1,\cdots,u_n)x(u_1)\cdots x(u_n)du^n\right)\\ &= \int_{R^{n+1}} F(u_1,\cdots,u_n)x(u_1)\cdots x(u_n)du^n\delta_t(u_{n+1})x(u_{n+1})du_{n+1}\\ &= \int_{R^{n+1}} \tilde{F}(u_1,\cdots,u_{n+1})x(u_1)\cdots x(u_n)x(u_{n+1})du^{n+1},\end{aligned}$$

where $\tilde{F}$ is the symmetrization of $F(u_1,\cdots,u_n)\delta_t(u_{n+1})$.

By taking its S-transform the assertion follows.

■

In view of the assertions given so far, we consider ∂_t, and ∂_t^* in parallel, being called an *annihilation operator* and a *creation operator*.

Theorem 4.2 *The following commutation relations hold:*

$$[\partial_t, \partial_s^*] = \delta(t-s).$$

The proof can be done in the same way as Proposition 1.7.

■

There are many operators acting on (L^2) that are related to the differential operators, however to discuss them, we need to introduce a space of suitable white noise functionals so that we shall come to this topic later.

4.2 Application to stochastic differential equation

It is one of the motivations of white noise analysis to establish a method of solving a stochastic differential equations by a similar, but generalized method of solving ordinary (non-random) differential equations. This motivation may be realized in some particular cases as are illustrated below.

Consider the stochastic differential equation

$$dX(t,\omega) = a(t,X(t,\omega))dt + b(t,X(t,\omega)dB(t,\omega),\ X(0,\omega) = 0. \quad (4.2.1)$$

Assume that $a(\tau,x)$ and $b(\tau,x)$

i) are continuous in τ.

ii) satisfy the Lipschitz condition : i.e. there exist constants k_1 and k_2 such that for all $\tau \in [0, T]$,

$$|a(\tau, x) - a(\tau, y)| \leq k_1|x - y| \quad \text{and} \quad |b(\tau, x) - b(\tau, y)| \leq k_2|x - y|$$

holds.

Under the above assumption, there exists a unique solution $\{X(t), t \in T\}$ which is a Markov process to the stochastic differential equation (ref. Itô and McKean[67]).

A general theory of such a stochastic differential equation can be developed by using stochastic integral due to K. Itô, which will be discussed in Section 5.3.

In what follows, we shall explain that for some cases we can efficiently use the S-transform to solve a stochastic differential equation, by examples.

By using the S-transform to the stochastic differential equation (4.2.1) to have non-random functional differential equation of $S(X(t))(\xi) = U(t, \xi)$, where the notations are modified suitably.

We assume that a and b are linear functional of $X(t)$. Then we have

$$S(a(t, X(t)))(\xi) = \tilde{a}(t, U(t, \xi)),$$

and similarly for $\tilde{b}(t, X(t))$. Hence, we have

$$\frac{d}{dt}U(t, \xi) = \tilde{a}(t, U(t, \xi)) + \tilde{b}(t, U(t, \xi)), \quad U(0, \xi) = 0.$$

Suppose this equation can be solved. Then applying S^{-1} transform to the solution of non-random differential equation, the required solution can be obtained.

Example 4.5 We now take a Langevin stochastic differential equation

$$dX(t) = -\lambda X(t)dt + dB(t),\ t \in (-\infty, \infty),\ \lambda > 0. \tag{4.2.2}$$

We first take S transform to have non-random differential equation

$$\frac{d}{dt}U(\xi) = -\lambda U(\xi) + \xi(t),$$

and then for fixed ξ, it is viewed as a linear ordinary differential equation

$$\frac{d}{dt}U = -\lambda U + \xi.$$

So the solution of which is

$$U(\xi) = e^{-\lambda t} \int_{-\infty}^{t} e^{\lambda u} \xi(u) du.$$

By taking S^{-1}-transform, the solution of (4.2.2) is

$$X(t) = e^{-\lambda t} \int_{-\infty}^{t} e^{\lambda u} \dot{B}(u) du.$$

Example 4.6 Consider a stochastic differential equation

$$dX(t) = aX(t)dt + (bX(t) + c)dB(t), \quad t \in [0, \infty), \tag{4.2.3}$$

with initial condition $X(0) = 0$.

Note that $dB(t) = B(t + dt) - B(t)$, so that it is independent of $X(t)$. In terms of white noise analysis terminology and by (4.2.3),

$$X(t)\frac{dB(t)}{dt} = \partial_t^* X(t).$$

Thus, equation (4.2.3) is expressed in the form

$$\frac{d}{dt}X(t) = aX(t) + \partial_t^*(bX(t) + c), \ t \in [0, \infty), \tag{4.2.4}$$

where $X(t)$ is assumed to be in $(L^2)^-$.

Apply the S-transform, we have

$$\frac{d}{dt}U(t, \xi) = aU(t, \xi) + \xi(t)(bU(t, \xi) + c),$$

where $U(t, \xi) = (SX(t))(\xi)$.

Once ξ is fixed, it is viewed as a linear ordinary differential equation

$$\frac{d}{dt}U - (a + b\xi)U - c\xi = 0. \tag{4.2.5}$$

Since $X(0) = 0$, we have $U(\xi, 0) = 0$, thus the solution of (4.2.5) is

$$U(\xi, t) = e^{\int_0^t (a+b\xi(s))ds} \int_0^t c\xi(s) e^{-\int_0^s (a+b\xi(u))du} ds.$$

Obviously, this functional $U(\xi, t)$ satisfies the Potthoff-Streit criterion, so that

$$S^{-1}U(t, \xi) = X(t)$$

exists and is the solution of equation (4.2.3). Exact form of the n-th component $X_n(t)$ of $X(t)$ is given by Hida[27], Section 4.6.

Example 4.7 Doob's Classical approach to a strictly multiple Markov Gaussian processes. Let L_t be an ordinary operator of order N. Consider a stochastic differential equation

$$L_t X(t) = \dot{B}(t), \ t \geq 0, \tag{4.2.6}$$

with the initial conditions

$$X(0) = X'(0) = \cdots = X^{(N-1)}(0) = 0,$$

where L_t is an ordinary differential operator of order N.

The equation was considered to be a formal expression, since $\dot{B}(t)$ is not an ordinary random variable. The expression (4.2.6) is acceptable in $\mathcal{H}_1^{(-1)}$.

We have the space of generalized random variables which are linear in $\dot{B}(t)$.

It is known that the solution is expressed as

$$X(t) = \int_0^t R(t, u)\dot{B}(u)du, \tag{4.2.7}$$

where $R(t, u)$ is the Riemann function associated with L_t. (See Hida[21].)

What have been discussed are as follows. Assume that the $X(t)$ is $(N-1)$-times differentiable in the strong topology in $\mathcal{H}_1$, and there exists an $(N-1)$th ordinary differential operator $L_t^{(1)}$ such that

$$L_t = \frac{1}{v_0(t)} \frac{d}{dt} \frac{1}{v_1(t)} L_t^{(1)}$$

and

$$L_t^{(1)} X(t) = Y(t), \ Y(0) = 0. \tag{4.2.8}$$

The solution $Y(t)$ is a simple Markov and it satisfies a stochastic differential equation

$$\frac{d}{dt}\frac{1}{v_1(t)}Y(t) = v_0(t)dB(t),$$

where v_0 and v_1 are smooth functions which may be assumed being never vanish over $[0,\infty)$.

The $Y(t)$ is expressed as

$$Y(t) = \int_0^t v_1(t)v_0(u)\dot{B}(u)du.$$

Since $Y(t)$ is continuous, so that we can solve equation (4.2.8) easily and the solution (4.2.7) is obtained.

One may ask how to get the operator $L_t^{(1)}$ in equation (4.2.8). The answer comes from the Frobenius expression of the operator L_t in such a way that

$$L_t = \frac{1}{v_0(t)}\frac{d}{dt}\frac{1}{v_1(t)}\frac{d}{dt}\cdots\frac{d}{dt}\frac{1}{v_N(t)},$$

where $v_i(t), i = 0, 1, \cdots, N$, are determined by the fundamental solutions of the linear differential equation

$$L_t v(t) = 0.$$

We can give more direct method to obtain the solution of (4.2.6) by using the S-transform.

Set $(SX(t))(\xi) = U(t,\xi)$. By assumption on $X(t)$ we can prove that $U(t,\xi)$ is differentiable in t as many times as N. Hence, equation (4.2.6) turns out to be

$$L_t U(t,\xi) = \xi(t),$$

where $U^{(k)}(0,\xi) = 0, k = 0, 1, \cdots, N-1$, with $U^{(k)}(t,\xi) = \frac{d^k}{dt^k}U(t,\xi)$.

The solution is

$$U(t,\xi) = \int_0^t R(t,u)\xi(u)du.$$

The $S^{-1}(U(t,\xi))$ gives (4.2.7), the solution of (4.2.6).

4.3 Differential calculus and Laplacian operators

Set

$$\Delta_\infty = \int \partial_t^* \partial_t dt. \tag{4.3.1}$$

Proposition 4.6 *For φ in $\mathcal{H}_n, n \geq 0$, we have*

$$\Delta_\infty \varphi = n\varphi. \tag{4.3.2}$$

Proof. For $\varphi \in \mathcal{H}_n^{(n)}$ with kernel function $F(u_1, \cdots, u_n)$, $\partial_t \varphi$ has the kernel $nF(u_1, \cdots, u_{n-1}, t)$. Then apply ∂_t^* to have the multiplication by $\delta_t(u_n)$. Hence the integration with respect to dt proves the assertion.

■

This results shows that Δ_∞ is densely defined on (L^2).

Definition 4.5 In view of the above proposition, Δ_∞ is called the **number operator**. Sometimes $-\Delta_\infty$ is called the *infinite dimensional Laplace-Beltrami operator.*

As is expected, the operator $-\Delta_\infty$ is negative:

$$\begin{aligned}\langle -\Delta_\infty \varphi, \varphi \rangle &= -\int \langle \partial_t^* \partial_t \varphi, \varphi \rangle dt \\ &= -\int \|\partial_t \varphi\|^2 dt \leq 0.\end{aligned}$$

As a finite dimensional analogue, we claim

Theorem 4.3 *The operator $\partial_t + \partial_t^*$ is a multiplicative operator p_t (= $\dot{B}(t)\cdot$) from $(L^2)^+$ to $(L^2)^-$:*

$$p_t = \partial_t^* + \partial_t. \tag{4.3.3}$$

Proof. We use the property of Hermite polynomial,

$$H_{n+1}(x) - \frac{x}{n+1} H_n(x) + \frac{\sigma^2}{n+1} H_{n-1}(x) = 0, \ n \geq 1,$$

for Hermite polynomials $H_n(x) = H_n(x, \sigma^2)$ with parameter σ^2. (See the appendix.)

Taking $x = \frac{\Delta B}{\Delta}$and $\sigma^2 = \frac{1}{\Delta}$, accordingly we have

$$H_{n+1}\left(\frac{\Delta B}{\Delta};\frac{1}{\Delta}\right)-\frac{1}{n+1}\frac{\Delta B}{\Delta}H_n\left(\frac{\Delta B}{\Delta};\frac{1}{\Delta}\right)+\frac{1}{n+1}\frac{1}{\Delta}H_{n-1}\left(\frac{\Delta B}{\Delta};\frac{1}{\Delta}\right)=0. \tag{4.3.4}$$

We know that by letting $\Delta \to \{t\}$ the limit of each term exists in the space $(L^2)^-$. Hence we have

$$\dot{B}(t)H_n\left(\dot{B}(t);\frac{1}{dt}\right)=H_{n-1}\left(\dot{B}(t);\frac{1}{dt}\right)\frac{1}{dt}+(n+1)H_{n+1}\left(\dot{B}(t);\frac{1}{dt}\right).$$

Namely, we have

$$\dot{B}(t)\cdot :\dot{B}(t)^n :=: (\dot{B}(t))^{n+1} : +n : (\dot{B}(t))^{n-1} : \frac{1}{dt}.$$

We can therefore define the multiplication by p_t $(= \dot{B}(t)\cdot)$ using the above formula. Thus, we have

$$p_t : \dot{B}(t)^n :=: \dot{B}(t)^{n+1} : +n : \dot{B}(t)^{n-1} : \frac{1}{dt}.$$

For the Hermite polynomials in $\dot{B}(t)$ and hence for any Hermite polynomials $\varphi(\dot{B})$ in $\dot{B}(t)$'s, we have

$$p_t\varphi = \partial_t^*\varphi + \partial_t\varphi.$$

This relationship can be generalized to $\mathcal{H}_n^{(-n)}$, and eventually this is true for all the $\varphi \in (L^2)^-$, so far as ∂_t^* and ∂_t are defined.

■

Remark 4.2 *Compare the computation of multiplication done in the discrete case and see the passage in Section 2.3.*

By using the formula (4.3.3), the operator $-\Delta_\infty$ is written as

$$-\Delta_\infty = -\int (p_t - \partial_t)\partial_t dt = \int (\partial_t^2 - p_t\partial_t)dt.$$

With this expression $-\Delta_\infty$ is viewed as an infinite dimensional analogue of the finite dimensional spherical Laplacian or of the Laplace-Beltrami operator acting on $L^2(S^n, d\theta)$, $d\theta$ being the uniform measure on S^n.

Example 4.8 Easy computations prove that

$$p_t\dot{B}(s) =: \dot{B}(t)\dot{B}(s) : +\delta(t-s).$$

Another finite dimensional analogue is the *rotation*

$$r_{t,s} = p_t \partial_s ds - p_s \partial_t dt.$$

Example 4.9 As is expected, we have

$$\gamma_{t,s} \dot{B}(t) = -\dot{B}(s).$$

Proposition 4.7 *The infinite dimensional Laplace-Beltrami operator commutes with all the rotations.*

Proof. Actual computations show that

$$[-\Delta_\infty, r_{t,s}] = 0 \quad \text{for any } t, s,$$

where $[\cdot, \cdot]$ is the Lie bracket. ∎

Remark 4.3 *The following properties are to describe the connection between Laplace-Beltrami operator Δ_∞ and Hida quadratic functionals, i.e. with the choice $n = 2$.*

i) $\mathcal{H}_2$ *is the eigenspace of* Δ_∞ *with eigenvalue 2.*

ii) $\mathcal{H}_2^{(-2,1)}$ *(which will be defined in Chapter 5) is also eigenspace of* Δ_∞.

iii) $\int g(u)\dot{B}(u)\dot{B}(h(u))du$, *$h$ being general and $h(u) \neq u$, is not always an eigenfunctional.*

We may consider another Laplacian Δ_V called the *Volterra Laplacian* or Gross Laplacian given by

$$\Delta_V = \int \partial_t^2 dt. \tag{4.3.5}$$

It plays some roles in white noise analysis, and as we shall discuss later, an interesting example of a quadratic form of ∂_t. The domain of Δ_V is wide enough including $\sum_n \mathcal{H}_n^{(n)}$.

Then, we come to a second order differential operator denoted by Δ_L which will be called later the *Lévy Laplacian*, given by

$$\Delta_L = \int \partial_t^2 (dt)^2. \tag{4.3.6}$$

Concerning the operator Δ_L, it is necessary to have further consideration before giving the name.

Note that such a conventional notation comes from the Volterra-Lévy principle "le passage du fini al'infini" in the case where we have continuously many vectors, that is our case in question $\{\dot{B}(t), t \in R\}$ as was discussed in Sections 2.3 and 2.4.

The notation $(dt)^2$ in the above expression seems strange. But we can give plausible reasons for using such a notation. One dt is for the integration and another dt, formally speaking, is used to cancel $\frac{1}{dt}$ arising from the integrand because of the effect of applying ∂_t^2. Some more interpretations can be seen from the examples that will come later.

On Laplacian

It is impossible to state a general way of understanding the Laplacian, however we wish to think of essentials of the notion of Laplacian in analysis. Still, it is difficult to consider a general way of defining. Here, we shall take two ways of understanding a Laplacian.

First, we remind two different aspects of the white noise measure space (E^*, μ). It is impossible to define the "support" of the measure μ, however we may estimate the set where μ is supported.

(1) As we have seen before, the strong law of large numbers suggests us to consider μ like a uniform measure on an infinite dimensional sphere with radius $\sqrt{\infty}$. Moreover, it looks like, as it were, a symmetric space defined by the infinite dimensional rotation group. To be more precise, the group G_∞ in Hida's favorite picture (Fig. 4.1) characterizes Δ_∞, which we call the infinite dimensional Laplace-Beltrami operator in such a way that Δ_∞, is a negative quadratic form of the Lie algebra of G_∞.

    The Fock space provides a decomposition of the unitary representation of G_∞. These facts are to be very much satisfactory.

    Note that the story mentioned just above is closed with the G_∞, not in entire group $O(E) = O_\infty$.

(2) Another aspect is that μ is supported by a "linear" space. The μ is quasi-invariant under the translation by $L^2(R^1)$-functions, unfortunately not translation invariant. We can, therefore, think of the similar

form to the finite dimensional case. Namely, the sum of the second order differential operators; the sum of ∂_t^2 to have $\int \partial_t^2 (dt)^2$. Here $(dt)^2$ is taken instead of dt, as is explained above.

Then, naturally, harmonic properties is claimed in connection with the infinite dimensional rotation group. In this case, whiskers, see Fig. 4.1, play more significant roles.

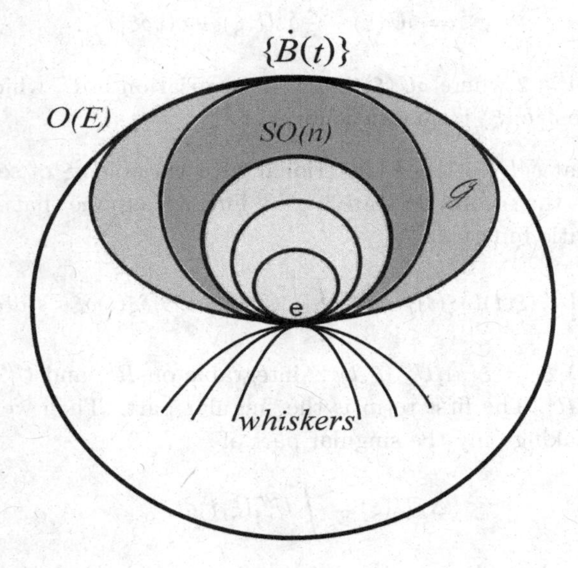

Fig. 4.1

We should recognize the characteristics of the expression of Δ_L using $(dt)^2$, the use of S-transform and the renormalization of quadratic terms of $\dot{B}(t)$'s. The passage from discrete expression of Δ_L to continuous case tells us the basic property of $\dot{B}(t)$ with scale $\frac{1}{\sqrt{dt}}$. The S-transform is efficient enough to make such unfamiliar properties be clear in the ordinary language of calculus.

So far we have discussed particular second order differential operators, in particular Laplacians. We now come to a general formula of the second order differential operators.

As in the case of ∂_t, we define the second order variation and differential operators with the help of the S-transform, i.e. U-functional.

Let $U(\xi), \xi \in E$, be a U-functional, i.e. $U(\xi) = (S\varphi)(\xi)$ for some $\varphi \in (L^2)^-$. We take the second variation $\delta^2 U$ in the sense of Fréchet (not of Gâteaux), such that

$$\Delta U(\xi) \equiv U(\xi + \delta\xi) - U(\xi)$$
$$= \delta U(\xi) + \frac{1}{2}\delta^2 U(\xi) + o(\|\delta\xi\|^2),$$

for some $\| \ \|, n \geq 2$ where $\delta U(\xi)$ is the first variation of U, which we have discussed. The $\delta^2 U(\xi)$ is quadratic form of $\delta\xi$.

Assume that $\delta^2 U(\xi)$ is a U-functional with variable $\delta\xi$ of some $\mathcal{H}_2^{(-2)}$-functional ψ is the second derivative of φ. Further assume that $\delta^2 U(\xi)$ is a normal quadratic functional.

$$\delta^2 U(\xi) = \int U''_{\xi^2}(\xi, t)(\delta\xi(t))^2 dt + \int\int U''_{\xi\xi_1}(\xi, t, s)\delta\xi(t)\delta\xi_1(s)dtds, \quad (4.3.7)$$

where $\xi = \xi(t), \xi_1 = \xi(s), U''_{\xi^2}(\xi, t)$ is integrable on R^1, and $U''_{\xi\xi_1}(\xi, t, s)$ is integrable on R^2. The first term is the singular part. Then we define the operator $\tilde{\Delta}_L$ taking only the singular part of

$$\tilde{\Delta}_L U(\xi) = \int U''_{\xi\xi}(\xi, t)dt. \quad (4.3.8)$$

Note that we have discussed, in Section 3.4, a quadratic form by dividing into two parts, and we know the two parts should be discriminated. The first term in (4.3.7) is the singular term, while the second term is the regular term (cf. Q_1 and Q_2 in Section 3.5). The $(\delta\xi(t))^2$ in the first term is an infinite dimensional analogue of $\sum(dx_i)^2$.

Example 4.10 Take a normal form of quadratic functional, like

$$U_2(\xi) = \int f(t)\xi(t)^2 dt.$$

The Lapalcian is to be twice of the trace of the second variation. In this case $2\int f(t)dt$. Now we have

$$\delta U_2(\xi) = 2\int f(t)\xi(t)\delta\xi(t)dt.$$

That is

$$U_2'(\xi)(t) = 2f(t)\xi(t) = \int 2f(t)\xi(u)\delta_t(u)du,$$

the variation of which is

$$\delta U_2'(\xi) = 2f(t)\delta_t(t) = 2f(t)\frac{1}{dt}.$$

To have the trace we need an additional dt to cancel $\frac{1}{dt}$, namely

$$2f(t)\frac{1}{dt}\cdot(dt)^2$$

is the integrand to have

$$\tilde{\Delta}_L U = 2\int f(t)dt.$$

At this stage some relations between the expressions of Δ_L in the discrete parameter case and in the $L^2[0,1]$ continuous case, as above, may be seen by taking the parameter set to be $[0,1]$ and the choice of the complete orthonormal system $\{\xi_n\}$ to be that of an equally dense system, namely

$$\lim \frac{1}{N}\sum_1^N \xi_n(u)^2 = 1$$

and

$$\lim \frac{1}{N}\sum_1^N \xi_n(u)\xi_n(v) = 0, a.e.$$

We are now ready to define the Lévy Laplacian. Note that $\mathcal{F}$ is the image of $(L^2)^-$ under the S-transform, and let $\mathcal{D}$ be a subspace of $\mathcal{F}$ containing $U(\xi)$ satisfying the following conditions :

i) U is second order Fréchet differentiable,
ii) $\delta^2U(\xi)$ is a U-functional and is expressed in the form (4.3.7) with $U''_{\xi^2}(\xi,\cdot)\in L^1(R^1)$ and with $U''_{\xi\xi_1}(\xi,\cdot,\cdot)\in L^2(R^2)$, for every ξ.

Define $D(\Delta_L)$ by

$$D(\Delta_L) = \{\varphi; S\varphi \in \mathcal{D}\}.$$

Let $\tilde{\Delta}_L$ be the trace of second order functional derivative such that the domain of $\tilde{\Delta}_L$ is in $\mathcal{D}$ and that

$$\tilde{\Delta}_L U(\xi) = \int U_{\xi^2}(\xi, t)dt.$$

Definition 4.6 Let Δ_L be a quadratic form of ∂_t's such that

i) Domain of Δ_L is $D(\Delta_L)$

ii) $\Delta_L \varphi = S^{-1}(\tilde{\Delta}_L U)$, where $U = S\varphi$, with $\varphi \in D(\Delta_L)$.

Then, Δ_L is called the Lévy Laplacian.

Proposition 4.8

i) If $\varphi \in (L^2)$, then $\Delta_L \varphi = 0$.

ii) The Lévy Laplacian commutes with the rotation $r_{s,t}$.

Proof. (i) As is easily seen, Δ_L annihilates all the (L^2)-functional. The closure of the domain of Δ_L includes (L^2).

(ii) The commutation

$$\begin{aligned}[\Delta_L, r_{s,t}] &= \left[\int \partial_t^2 (dt)^2, \partial_s^* \partial_u - \partial_u^* \partial_s\right] \\ &= \int \partial_t^2 (dt)^2 \partial_s^* \partial_u - \int \partial_t^2 (dt)^2 \partial_u^* \partial_s - \partial_s^* \partial_u \int \partial_t^2 (dt)^2 + \partial_u^* \partial_s \int \partial_t^2 (dt)^2 \\ &= 0.\end{aligned}$$

■

Remark 4.4 *We may consider a general Lie algebra generated by ∂_t and $\partial_s^*, t, s \in R^1$. As we observed one of the significant quadratic forms is the Lévy Laplacian which commutes with the rotations.*

Lie algebras generated by the polynomials in ∂_t, ∂_s^* of degree higher than 2 should include polynomials with much higher degree, i.e. algebras closed under the Lie product will include polynomials, the degrees of which are unbounded.

Example 4.11 The Gauss kernel is given by

$$\varphi_c = N e^{c \int x(t)^2 dt}, \tag{4.3.9}$$

where the constant N is the renormalizing constant. It is determined to adjoint the S-transform $U(\xi)$ given below.

The function φ_c is in $D(\Delta_L)$ and is an eigen-functional of Δ_L. In fact

$$\Delta_L \varphi_c = \frac{2c}{1-2c}\varphi_c. \tag{4.3.10}$$

For proof, we see that the U-functional is

$$U(\xi) = (S\varphi)(\xi) = \exp\left[\frac{c}{1-2c}\int \xi(u)^2 du\right],$$

and then its variation is

$$\delta U(\xi) = U(\xi)\int \frac{2c}{1-2c}\xi(u)\delta\xi(u)du, \tag{4.3.11}$$

and the first Fréchet derivative is

$$U'(\xi) = \frac{2c}{1-2c}U(\xi)\xi(t), \tag{4.3.12}$$

the variation of which is

$$\delta U'(\xi,t) = \frac{2c}{1-2c}U(\xi)\delta\xi(t) + (\frac{2c}{1-2c})^2 U(\xi)\xi(t)\int \xi(u)\delta\xi(u)du.$$

In other words, the second Fréchet derivative is

$$U''(\xi,t,s) = \frac{2c}{1-2c}U(\xi)\delta(t-s) + (\frac{2c}{1-2c})^2 U(\xi)\xi(t)\xi(s).$$

Thus,

$$U''_{\xi\xi}(\xi,t,s) = \frac{2c}{1-2c}U(\xi).$$

Hence, we have

$$\tilde{\Delta}_L U = \int U''_{\xi\xi'}(\xi,t)dt = \int \frac{2c}{1-2c}U(\xi)dt,$$

on the parameter set [0,1]. Accordingly,

$$\Delta_L \varphi = \frac{2c}{1-2c}\varphi.$$

Then, we can prove a surprising properties of Δ_L.

Proposition 4.9 *$\tilde{\Delta}_L$ is a derivation.*

Proof. The second variation $\delta^2(U \cdot V)$ is the sum

$$\delta^2(U \cdot V) = U \cdot \delta^2 V + 2\delta U \cdot \delta V + \delta^2 U \cdot V.$$

We assume that both $\delta^2 U$ and $\delta^2 V$ have normal forms. The first term and the third term involve singular factors, but the second term does not have singular factor. Thus, we take the singular factors of the first term and the third term. Thus, according to the definition of $\tilde{\Delta}_L$ given in (4.3.8), we have

$$\tilde{\Delta}_L(U \cdot V) = U\tilde{\Delta}(V) + \tilde{\Delta}U \cdot V.$$

Hence the assertion follows.

■

Remark 4.5 *Unlike the Laplace-Beltrami operator Δ_∞, we now give the connection between Lévy Laplacian Δ_L and Hida quadratic functionals.*

i) Δ_L acts on $\mathcal{H}_2^{(-2)}$ and annihilates $\mathcal{H}_2$.

ii) For $\varphi = \int f(u) : \dot{B}(u)^2 : du$, $\Delta_L \varphi = 2 \int f(u) du$.

iii) $\int g(u)\dot{B}(u)\dot{B}(h(u))du$, with $h(u) \neq u$ for every u, is not in the domain of Δ_L.

4.4 Infinite dimensional rotation group $O(E)$

Apart from the Hida calculus, this section is devoted to introducing the infinite dimensional rotation group in continuous parameter case. It has more complex structure than the group introduced in Chapter 1, since the parameter is continuous.

Definition 4.7 Let E be a suitable nuclear space. A transformation g on $L^2(R)$ is a *rotation* of E if it satisfies the following.

i) g is a linear transformation on E such that $\|gf\| = \|f\|, f \in L^2(R)$, (orthogonal transformation) and
ii) g is a homeomorphism of E.

Let $O(E)$ be the collection of all rotations of E. The product of rotations is defined in the usual manner such that for g_1, g_2 in $O(E)$,

$$(g_1 g_2)\xi = g_1(g_2\xi), \quad \xi \in E. \tag{4.4.1}$$

Since a rotation g is isomorphism of E, the inverse g^{-1} is defined. Thus, the $O(E)$ forms a group. It is natural to introduce the compact open topology since it is a transformation group.

Definition 4.8 The topologized group $O(E)$ is called a rotation group of E, or if E is not specified, it is simply called an *infinite dimensional rotation group* and is denoted by $O(E)$.

For any $g \in O(E)$ and fixed $x \in E^*$, the mapping

$$\xi \mapsto \langle x, g\xi \rangle$$

is a continuous linear functional on E, according to Definition 4.7. So there exists $g^*x \in E^*$ such that

$$\langle x, g\xi \rangle = \langle g^*x, \xi \rangle.$$

Thus, we have a mapping :

$$x \mapsto g^*x$$

which defines a linear automorphism g^* of E^*. Let

$$O^*(E^*) = \{g^*, g \in O(E)\}.$$

Then, $O^*(E^*)$ forms a group with the product

$$(g_1^* g_2^*)x = g_1^*(g_2^* x), \quad g_1^*, g_2^* \in O^*(E^*).$$

Note that $(g^*)^{-1} = (g^{-1})^*$.

Define the product $g^* \cdot \mu$ by

$$(g^* \cdot \mu)(B) = \mu(g^* B), \quad B \in \mathcal{B}.$$

Theorem 4.4 *Each g^* is a μ measure preserving transformation acting on (E^*, μ). That is, μ is $O^*(E^*)$-invariant.*

Proof. We have to prove that $g^* \cdot \mu = \mu$, for all $g^* \in O^*(E^*)$. The proof is similar to the discrete case. Actually the characteristic functional of the measure $g^* \cdot \mu$ is

$$\begin{aligned}\int_{E^*} e^{i\langle x,\xi\rangle} d(g^* \cdot \mu)(x) &= \int_{E^*} e^{i\langle x,\xi\rangle} d\mu(g^* x) \\ &= e^{-\frac{1}{2}\|g^{-1}\xi\|^2} \\ &= e^{-\frac{1}{2}\|\xi\|^2}.\end{aligned}$$

■

Remark 4.6 *White noise analysis has an aspect of the harmonic analysis arising from the infinite dimensional rotation group. The harmonic analysis can be, in some parts, approximated by finite dimensional analysis. But, to be very important, there are lots of significant results that are essentially infinite dimensional; in fact, those results cannot be well approximated by finite dimensional concepts.*

We have given the definition of the shift in Section 3.4. The adjoint of S_t is denoted by T_t. The $\{T_t\}$ is a one-parameter group of measure μ preserving transformations. Namely, $\{T_t\}$ is a *flow* on (E^*, μ).

Thus, the unitary operator $U_t : (U_t\varphi)(x) = \varphi(T_t x)$ is defined, and if it is restricted to $\mathcal{H}_1$, it coincides with U_t defined in (3.3.4).

We now think of the connection with the quadratic functionals discussed in Section 3.4. We see that $\{U_t\}$ has the unit multiplicity on the space $\{Q_1\}$, while on $\{Q_2\}$ it has countably infinite spectrum, where Q_1 and Q_2 are defined in Section 3.5.

The unitary operator U_t is a continuous representation of the space spanned by $\{Q_1, Q_2\}$. On the spaces spanned by $\{Q_1\}$ or $\{Q_2\}$, $\{U_t\}$ has a continuous spectrum. Thus, we can discriminate $\{Q_1\}$ and $\{Q_2\}$ by the unitary representation of the shift. We now change back the notation from x to $\dot{B}$. The space spanned by $Q_2(\dot{B})$ is in $\mathcal{H}_2^{(-2)}$ on which we have a unitary representation U_g in such a way that

$$U_g : \varphi \mapsto (U_g\varphi)(\dot{B}) = \varphi(g^*\dot{B}), \quad \varphi \in \mathcal{H}_2^{(-2)}.$$

We now discuss another intensity property of quadratic functionals in connection with the rotation group $O(E)$.

The renormalized limit of Q_1 satisfies certain invariance. The collection of such limits accepts an irreducible continuous representation of the group G the collection of the 2×2 matrices of the form

$$\begin{pmatrix} a & b \\ 0 & 1 \end{pmatrix}$$

where $a \neq 0, b \in R^1$.

Define a subspace L_2^* of $\mathcal{H}_2^{(-2)}$ by

$$L_2^* = \left\{ \int h(u) : \dot{B}(u)^2 : du \ ; \ h \in K^{-1}(R^1) \right\}.$$

We use the terminology x, the sample function of $\dot{B}$, instead of $\dot{B}(t)$ in the following.

Proposition 4.10 *An irreducible continuous representation of the group G is given on the space L_2^* in such a way that for $g \in G$:*

$$g : u \ \mapsto \ au + b. \tag{4.4.2}$$

Then,

$$U_g \varphi(x) = \int h(au + b) : x(u) :^2 du.$$

Proof. Since φ is in L_2^*, it is expressed in the form

$$\varphi(\dot{B}) = \int h(u) : x(u)^2 : du, \quad h \in K^{-1}(R^1).$$

From (4.4.2), we have $g : \xi(u) \ \longmapsto \ \xi(au + b)\sqrt{|a|}$.

$$\begin{aligned} \langle \xi, g^* x \rangle &= \langle g\xi, x \rangle \\ &= \int \xi(au + b)\sqrt{|a|} x(u) du \\ &= \int \xi(v) x(\frac{v - b}{a}) \frac{1}{\sqrt{|a|}} dv. \end{aligned}$$

Thus, we have

$$g^* x = x(\frac{v - b}{a}) \frac{1}{\sqrt{|a|}}.$$

$$
\begin{aligned}
U_g\varphi(x) &= \varphi(g^*x) \\
&= \int h(v) : (g^*x)^2 : dv \\
&= \int h(v) : (x(\frac{v-b}{a}))^2 : \frac{1}{|a|} du \\
&= \int h(au+b) : x(u)^2 : du.
\end{aligned}
$$

The kernel function is an image of a $K^{-1}(R^1)$-continuous mapping of h under g. Irreducibility is implied from the unitary representation of the group G on $L^2(R^1)$.

■

4.5 Addenda

We have, so far, discussed unitary representation of a group in many cases. This section explains the general idea.

Suppose that $\mathcal{H}$ is a complex Hilbert space, that is either finite or infinite dimensional. Let G be a topological group on a Hilbert space $\mathcal{H}$.

Definition 4.9 A map U_g, associated to every member $g \in G$, that is a linear operator acting on $\mathcal{H}$ is called a *unitary representation* of G if the following conditions 1), 2) and 3) are satisfied.

1) Each U_g is unitary,
2) $g \mapsto U_g$ is a homomorphism $G \mapsto \mathcal{U}$, where $\mathcal{U} = \{\text{unitary operators on } \mathcal{H}\}$,
3) $g \mapsto U_g\varphi$ is continuous from G to $\mathcal{H}$.

By definition

$$
U_{gg'} = U_g U_{g'}, \quad U_g^{-1} = U_{g^{-1}}
$$

holds.

A unitary representation is called finite or infinite dimensional according as the Hilbert space $\mathcal{H}$ is finite or infinite dimensional.

There is a particular unitary representation of G called a *regular* left (right) unitary representation of a locally compact topological group G. It is given in such a way that the Hilbert space is taken to be $\mathcal{H} = L^2(G, dm)$, and

$$U_g f(x) = f(g^{-1}x), \ x \in G, f \in \mathcal{H}.$$

Obviously this satisfies the conditions 1), 2) and 3), so that it defines a unitary representation of G. This representation is called the *left regular representation* of G.

The *right regular representation* V_g is defined by

$$V_g f(x) = f(xg), \quad g \in G, f \in \mathcal{H}.$$

Definition 4.10 A unitary representation is called *irreducible* if the Hilbert space $\mathcal{H}$ has no invariant subspace under all $U_g, g \in G$, except trivial subspace $\{0\}$ or the entire $\mathcal{H}$.

Take the unitary group E. We can form a unitary representation of $O(E)$ on the Hilbert space (L^2).

A function φ on $O(E)$ defined by

$$\varphi \mapsto (U_g\varphi)(x) = \varphi(g^*x), \ x \in E^*(\mu).$$

Then

$$\begin{aligned} \|U_g\varphi\|^2 &= \int |\varphi(g^*x)|^2 d\mu(x) \\ &= \int |\varphi(y)|^2 d\mu(y), \ y = g^*x \end{aligned}$$

since $d\mu(g^*x) = d\mu(x)$.

This shows that U_g is a unitary operator on (L^2). It is easy to show that

$$U_{g_1}U_{g_2} = U_{g_1g_2},$$

since $g_2^*(g_1^*x) = (g_1g_2)^*x$.

The example of U_g is described in the previous section.

Proposition 4.11 *There is a unitary representation of $O(E)$ on the Hilbert space (L^2).*

Chapter 5

Stochastic Integral

5.1 Introduction

Theory of stochastic integral has a long history. Under the title "Stochastic Integrals" there are many kinds of integrals, most of which are based on random measures. We are interested in Wiener integral, Multiple Wiener integrals, Itô integral and a generalization to Hitsuda-Skorokhod integrals. Furthermore, we can discuss stochastic integrals that use generalized random measure.

We now give a brief overview of various stochastic integrals.

As we shall see, in what follows, a stochastic integral is not quite an inverse operation of differentiation using the operator ∂_t, but a particular mapping of $(L^2)^-$ into itself.

First we define the Wiener integral expressed in the well-known form

$$X = \int f(u)dB(u), \tag{5.1.1}$$

where f is a square integrable non-random function and $dB(u)$ is a random measure.

Also we note that the multiple Wiener integral is not quite a stochastic version of the multiple integral of ordinary functions. It is a representation of the homogeneous chaos and can be investigated referring the properties of Hermite polynomials in Gaussian random variables.

We are familiar with the Itô stochastic integral. If the random measure is formed from a Brownian motion, we have a stochastic integral expressed in the form

$$\int \varphi(u)dB(u),\ \varphi(u) = \varphi(u,\omega), \tag{5.1.2}$$

where $\varphi(u)$ is a random function which is *non-anticipating* with respect to the sigma-field $\mathcal{B}_t(B)$ generated by the Brownian motion $B(s), s \le t$. The theory is beautiful and has lot of applications; for example in the theory of stochastic analysis and in other related fields.

We are going to discuss here another setup of stochastic integrals that extend not only the classical theory of stochastic integrals, but also give a new insight into the infinite dimensional integrations. In fact, we use the creation operators ∂_t^* that have been introduced in the previous chapter. As a result, the integral can naturally be extended to the Hitsuda-Skorokhod integrals.

Again, we should note that the white noise $\{\dot{B}(t), t \in R^1\}$ is taken to be the system of variables of generalized white noise functionals, in addition to the contribution to random measure as $\dot{B}(t)dt = dB(t)$. The members of the system are as many as *continuously many.* Each $\dot{B}(t)$ is a rigorously defined variable as a member of $\mathcal{H}_1^{(-1)}$.

The rigorous definition of the partial derivative, the so-called "Hida derivative", ∂_t has been given in (4.1.4). The adjoint operator ∂_t^* has also been defined in the usual manner, and actually it serves to define a generalization of the stochastic integral with respect to $\dot{B}(t)dt$ or $dB(t)$.

Thus, the Itô integral defined later in Section 5.3, expressed in the form (5.1.2), may be written as

$$\int \varphi(t)\dot{B}(t)dt.$$

Then, it is expressed in the form, by using the creation operator ∂_t^*,

$$\int \partial_t^* \varphi(t)dt. \tag{5.1.3}$$

This expression enables us to discuss the case where $\varphi(t)$ is not necessary to assume the non-anticipating condition. Hence the integrand of the left-hand side is considered as $\dot{B}(t)\varphi(t)dt$. In a general case, it is expressed as

$$\int (\partial_t + \partial_t^*)\varphi(t)dt,$$

where ∂_t and ∂_t^* are creation and annihilation operators, respectively.

As is mentioned above, we will deal with creation and annihilation operators to have a well understanding of Itô's stochastic integral.

Stochastic integrals that we are going to discuss, using creation operators, have the following meanings.

1) In general, the stochastic integrals are representations of random evolutional phenomena interfered by a noise. For example, they are solutions of stochastic differential equations.
2) Multiple Wiener integrals may be considered as Hermite polynomials in $\dot{B}(t)$'s, indexed by the continuous parameter t. There is introduced the chaos degree, and the collection of integrals forms a graded algebra interms of degree.
3) Calculus on chaos : The chaos degree becomes one degree higher by stochastic integration.
4) There is a random version of the area enclosed by a curve, like in the elementary calculus. It is defined by the stochastic integral and is called the stochastic area. It is a particular quadratic form and play particular roles in applications, in particular statistics.
5) We can construct spaces of stochastic integrals where representations of continuous groups; e.g. unitary representation of rotation group. This fact has connections with the theory of Lie group.

Among these items, there are some overlapping.

Remark 5.1 *We denote the standard Gaussian random measure either by $dB(t)$ or $\dot{B}(t)dt$. In fact, we prefer the latter. Concerning these notations, it is better to note that P. Lévy used to use $\xi\sqrt{dt}$, where ξ denotes a standard Gaussian random variable. He has also used the notation $dX(t)$. One might think that we have made mistake because of the notation $\xi\sqrt{dt}$; but not quite. Having been familiarized we think that it is a suitable notation.*

There is another kind of stochastic integral, due to P. Lévy. The idea is interesting, so we shall briefly discuss it in Section 5.5.

We shall discuss in Section 5.6 the so-called path integral or the Feynman integral. The measure to be involved is defined over the function space, which is assumed to be the collection of quantum mechanical trajectories.

5.2 Wiener integrals and multiple Wiener integrals

The Wiener integral

Good interpretation of Wiener integral can be found in N. Wiener[135]. We first recall the *Wiener integral* of the form

$$\int_a^b f(u)dB(u), \tag{5.2.1}$$

where f is non-random $L^2(R)$-function. It is defined in the following manner.

Since f is in $L^2([a,b])$, we can take a sequence of step functions $\{f_k\}$ which converges to f in $L^2([a,b])$. We define a step function f_n such that

$$f_n = \sum_{i=1}^{N} a_i \chi_{[t_{i-1},t_i]}, \tag{5.2.2}$$

corresponds to which we define

$$S_n = \sum_{i=1}^{n} a_i(B(t_i) - B(t_{i-1})). \tag{5.2.3}$$

It can be seen that S_n is a Gaussian random variable with mean 0 and variance

$$\begin{aligned} E(S_n^2) &= \sum_{i=1}^{n} a_i^2(t_i - t_{i-1}) \\ &= \int_a^b f_n(t)^2 dt. \end{aligned}$$

We can see that $\{f_n\}$ is a Cauchy sequence in $L^2([a,b])$, so is $\{S_n\}$ in the space (L^2). Here, the Wiener integral can be defined as

$$\int_a^b f(t)dB(t) = \lim_{n\to\infty} S_n,$$

in (L^2). Note that it is independent of the choice of the sequence f_n, that approximates f.

The Wiener integral (5.2.1) is a Gaussian random variable with mean 0 and variance $\int_a^b f(t)^2 dt$.

The Wiener integral $\int f(t)dB(t)$ is additive in $f \in L^2(R^1)$, so that we can prove that the collection $\{\int f(u)dB(u); f \in L^2(R^1)\}$ forms a Gaussian system. Further the closure of the system is a subspace of (L^2). It is the space $\mathcal{H}_1$ that we have established before.

Another way to define the Wiener integral is more simple, since we have some background on white noise.

Take a bilinear form $\langle \dot{B}, \xi \rangle$ where ξ is a test function in E. The bilinear form defines a Gaussian random variable on the white noise space (E^*, μ).

Let f be an $L^2(R^1)$-function. Then, we can find a sequence $\{\xi_n\}$ in E that converges to f in $L^2(R^1)$. Take a sequence $\langle \dot{B}, \xi_n \rangle$ which forms a Cauchy sequence in $L^2(E^*, \mu)$ as is easily proved. Then, the limit $\lim \langle \dot{B}, \xi_n \rangle$ exists (in the mean). It can be denoted by $\langle \dot{B}, f \rangle$, since the limit does not depend on the choice of the sequence ξ_n that approximates f. The limit can be expressed in the form

$$\langle \dot{B}, f \rangle = \int f(u)\dot{B}(u)du,$$

that coincides with the Wiener integral (5.2.1) almost surely.

Multiple Wiener integrals

The integral

$$\int_a^b \cdots \int_a^{u_3} \int_a^{u_2} F(u_1, \cdots, u_n)dB(u_1) \cdots dB(u_n), \tag{5.2.4}$$

where $F \in L^2([a,b]^n)$, may be written as

$$\frac{1}{n!} \int_a^b \cdots \int_a^b F(t_1, \cdots, t_n) : \dot{B}(t_1) \cdots \dot{B}(t_n) : dt^n, \tag{5.2.5}$$

where F is symmetric $L^2(R^n)$ function. It is known as *multiple Wiener-Itô integral* which can be defined as in the case of Wiener integral. Thus, we do not go into the details, however we shall explain to some extent. The kernel function that is the integrand $F(t_1, \cdots, t_n)$ can be approximated in $L^2(D)$ with $D = \{(u_1, u_2, \cdots, u_n); a < u_1 < u_2, < \cdots < u_n < b\}$ by elementary functions F_k of the form

$$F_k(u_1, u_2, \cdots, u_n) = \sum a_{j_1, j_2, \cdots, j_n} \prod_{j=1}^n \chi_{[c_j, d_j]}(u_j).$$

The set $\prod_j [c_j, d_j]$ is a subset of D, and they are non-overlapping sets, their sum of such elementary sets in D approximates D. The integral

$$X_k = \int \cdots \int F_k(u_1, u_2, \cdots, u_n) dB(t_1) \cdots dB(t_n),$$

is defined by

$$\sum a_{j_1, j_2, \cdots, j_n} \prod_{j=1}^{n} (B(d_j) - B(c_j))$$

which is the sum of orthogonal (in the space $L^2(\Omega)$) random variables. We therefore have

$$E|X_k|^2 = \sum |a_{j_1, j_2, \cdots, j_n}|^2 \prod_j |d_j - c_j| = \|F_k\|, \quad \| \ \| : L^2(\Omega)\text{-norm}.$$

We can take a sequence F_k such that $F_k \to F$ in $L^2(D)$. Obviously, the X_k defined by F_k forms a Cauchy sequence, so that the l·i·m X_k exists. The limit is independent of the choice of the sequence F_k. Hence the limit just obtained can simply be written as

$$\int_{t_{n-1}}^{b} \cdots \int_{t_1}^{t_2} \int_{a}^{t_1} F(u_1, \cdots, u_n) dB(u_1) \cdots dB(u_n).$$

The symmetrization of the kernel F defines the multiple Wiener integral

$$\int \cdots \int_{R^n} F(u_1, \cdots, u_n) dB(u_1) \cdots dB(u_n).$$

However, there is a crucial remark. Intuitively speaking, the diagonal part is not included in the above integration. Namely, the contribution to the integral of the part $\prod_j (dB(t_j))^{p_j}, p_j \geq 2$, is ignored. The reason can be seen from the following simple example.

Consider the example that renormalization of $\dot{B}(t)^2$ is

$$: \dot{B}(t)^2 := \dot{B}(t)^2 - \frac{1}{dt}$$

and the S-transform of which is

$$S(: \dot{B}(t)^2 :) = \xi(t)^2.$$

The annoying term $\frac{1}{dt}$ is removed, and the S-transform gives an ordinary function. Such kind of renormalization can be generalized to that we have discussed in Chapter 4. In this connection, it is a good time to see the Wick product.

The Wick product gives us a way to define the stochastic integral in general. Take the system of variables $\dot{B}(t)$'s, i.e. white noise instead of x_i's in the formulas (3.10.5) to (3.10.7), we have

$$
\begin{aligned}
:1: &= 1 \\
:\dot{B}(u): &= \dot{B}(u) \\
:\dot{B}(u_1)\cdots\dot{B}(u_n): &= :\dot{B}(u_1)\cdots\dot{B}(u_{n-1}):\dot{B}(u_n) \\
&\quad - \sum_{i=1}^{n-1}\delta(u_n-u_i):\dot{B}(u_1)\cdots\dot{B}(u_{i-1})\dot{B}(u_{i+1})\cdots\dot{B}(u_n):.
\end{aligned}
$$

Notation : We denote the Wick product $:\dot{B}(u_1)\cdots\dot{B}(u_n):$ by $:\dot{B}(t)^{\otimes n}:$ from now on.

We can see that the S-transform of the Wick product of $\dot{B}(t)$'s is simple products of $\xi(t)$'s.

Example 5.1 We can easily prove

$$
S(:\dot{B}(u_1)\cdots\dot{B}(u_n):)(\xi) = \prod_{u_j:\text{different}} \xi(u_i), \quad n = 1, 2, \cdots.
$$

Thus, we can define an integral in the form

$$
\begin{aligned}
S&\left(\int\cdots\int F(u_1,\cdots,u_n):\dot{B}(u_1)\cdots\dot{B}(u_n):du_1\cdots du_n\right) \\
&= \int\cdots\int F(u_1,\cdots,u_n)\xi(u_1)\cdots\xi(u_n)du_1\cdots du_n. \qquad (5.2.6)
\end{aligned}
$$

Theorem 5.1 *Using these notations a member in $\mathcal{H}_n$ is written as*

$$
\int\cdots\int F(u_1,\cdots,u_n):\dot{B}(u_1)\cdots\dot{B}(u_n):du_1\cdots du_n, \qquad (5.2.7)
$$

where F is in $\widehat{L^2(R^n)}$.

Thus, the *multiple Wiener-Itô integral* is defined as a white noise functional, the S-transform of which is given by (5.2.6). The kernel F can be taken to be a symmetric L^2-function so that uniqueness of the integral representation holds.

5.3 The Itô integral

Our aim is to define an integral of the form

$$X(t) = \int_0^t a(s, X(s))ds + \int_0^t b(s, X(s))dB(s) \tag{5.3.1}$$

which plays an essential role in the theory of stochastic differential equation.

As for the first term of the integral there is no problem. It is an ordinary integral with respect to the Lebesgue measure. However, both the integrand and $dB(s)$ in the second term are random variables, so that we only note that the convergence of the integral is in the almost sure sense or in the $L^2(\Omega, P)$ sense or others. In any case, there is no problem. The second integral is based on random measure $dB(s)$. The correct meaning of the integral should be made clear. We shall define the integral based on $dB(s)$ in Itô's sense.

Let $\mathcal{B}_t$ be the σ-field generated by $B(s), s \le t$. We now define the integral

$$\int_0^t Y(s, \omega)dB(s, \omega),$$

where $Y(s, \omega)$ is $\mathcal{B}_s$-measurable. The probability parameter ω is explicitly written to make sure that they are random variables.

We start with an approximation of the integral, actually we assume that the integrand is a step (random) function $Y^{(n)}(t, \omega)$ such that it is a constant random variable on Δ_k for a partition $\{\Delta_k = [t_{k-1}, t_k)\}$ of $[0, a]$ with $|\Delta_k| = 2^{-n}a$,

1) $Y^{(n)}(t, \omega)$ is expressed in the form

$$Y^{(n)}(t, \omega) = Y(t_{k-1}, \omega), \ t \in [t_{k-1}, t_k),$$

2) $Y^{(n)}(t, \omega)$ is $\mathcal{B}_{t_k}$-measurable and has finite variance, where we use the notation $\mathcal{B}_t$ to express the sigma-field generated by $\{B(s), s \le t\}$.

For such a step function $Y^{(n)}(t, \omega)$, we define the integral

$$\int_0^a Y^{(n)}(t, \omega)dB(t) = \sum_1^n Y^{(n)}(t_{k-1}, \omega)\Delta_k B.$$

This integral is denoted by $I_a(Y^{(n)})$.

Proposition 5.1 *The following equality holds.*

$$E((I_a(Y^{(n)})^2) = \int_0^a E(Y^{(n)}(t,\omega)^2)dt.$$

Proof. To make the notations simple, we omit the super index (n) and ω. Since

$$\begin{aligned}
E\left((Y_{k-1}\Delta_k B)\cdot(Y_{j-1}\Delta_j B)\right) &= E\left(E\left(Y_{j-1}\Delta_j B\cdot Y_{k-1}\Delta_k B)|\mathcal{B}_{t_{j-1}}\right)\right), k>j \\
&= E\left(Y_{j-1}\Delta_j B\left(E(Y_{k-1}\Delta_k B)|\mathcal{B}_{t_{j-1}}\right)\right), \\
&= E\left(Y_{j-1}\Delta_j B\cdot 0\right) \\
&= 0
\end{aligned}$$

holds, we have

$$\begin{aligned}
E(I^{(n)}(Y)^2) &= E|(\sum Y_{k-1}B_k)^2| \\
&= E|\sum_k Y_{k-1}^2 B_k^2 + 2\sum_{j<k} Y_{k-1}B_k\cdot Y_{j-1}B_j| \\
&= \sum_k E|Y_{k-1}^2||\Delta_k|.
\end{aligned}$$

The last equation tends to

$$\int_0^a E(Y^{(n)}(t)^2)dt.$$

Hence the assertion is proved. ∎

The next step is the generalization of the integrand. Set $\mathcal{B} = \bigvee_t \mathcal{B}_t$. We assume that the integrand $Y(t,\omega)$ is $\mathcal{B}$-adapted, that is $Y(t,\omega)$ is $\mathcal{B}_t$-measurable for every t. The collection of such processes $Y(t)$ is denoted by $\mathbf{Y}$. Set

$$L^2(\mathcal{B}) = \{Y(\cdot,\omega); Y(t,\omega) : \mathcal{B}_t - \text{measurable}, \int_0^a E(Y(t)^2)dt < \infty\}.$$

Then, we have

Theorem 5.2 *The following assertions hold.*

i) For every $Y(\cdot,\omega)\in L^2(\mathcal{B})$ the integral

$$I_a(Y) = \int_0^a Y(t,\omega)dB(t,\omega)$$

is well defined.

2) $I_a(Y)$ *is linear in* Y. *In addition, it is continuous in* Y *in the sense that*

$$\int_0^a |Y_n(t)|^2 dt \to 0 \Longrightarrow \sup_{0\le t\le a} |I_t(Y_n)| \to 0, \text{as } n \to \infty.$$

Proof. 1) Given $Y(t)$ in (L^2), it can be approximated by step functions $Y_n(t)$, as before, by taking a partition $\{\Delta_k\}$. By taking partitions finer, we are given a sequence $\{Y_n(t)\}$ which forms a Cauchy sequence in $L^2(\mathcal{B})$:

$$\int_0^a E((Y_n(t) - Y_m(t))^2 dt \to 0,$$

as $n, m \to \infty$. Hence, by Proposition 5.1, $I_a(Y^{(n)})$ forms a Cauchy sequence, so that the limit of the sequence $I_a(Y_n)$ exists. The limit may depend on the choice of the partition. In reality, we can show that the limit is independent of the choice of the partition, if the partition is getting uniformly finer as $n \to \infty$. For, if two sequences of partitions are given, then we can take a finer sequence of partitions, for which the limit of $I_a(Y^{(n)})$ is equal to each of the two.

Hence the limit may be written as

$$\int_0^a Y(s,\omega)dB(s,\omega).$$

2) Linearity of $I_a(Y^{(n)})$ is obvious and continuity of $I_a(Y^{(n)})$ in Y can be shown by what we have discussed in 1).

■

Remark 5.2 *The integrand* $Y(t,\omega)$ *is* $\mathcal{B}_t$*-measurable, so by understanding* $dB(t)$ *to be* $B(t+dt) - B(t)$ *with* $dt > 0$, *we see that the integrand and the* $dB(t)$ *are independent.*

Remark 5.3 *It is known that* $I_t(Y)$ *is continuous in* t *almost surely. For the proof, we need much preparation, so that we just refer to the literature*[64].

Example 5.2

$$\int_0^t B(s)dB(s) = \frac{1}{2}(B(t)^2 - t).$$

Example 5.3 Stochastic area. See Section 3.5.

Theorem 5.3 *(K. Itô) The integral (5.4.2) exists with probability 1, if $a(t,x)$ and $b(t,x)$ satisfy the Lipschitz conditions:*

i) $a(t,x)$ and $b(t,x)$ are continuous in t for each x,
ii) there exist α and β such that

$$|a(t,x)-a(t,y)| \le \alpha|x-y|, \quad |b(t,x)-b(t,y)| \le \beta|x-y|.$$

■

5.4 Hitsuda-Skorokhod integrals

We now propose a stochastic integral in a general framework within the theory of Hida distributions. It is a big advantage that the creation operator ∂_t^* plays a key role in what we propose. Our aim is to define an integral with respect to $dB(t) = \dot{B}(t)dt$ where the integrand is not necessarily non-anticipating.

Formally writing the integral of the form

$$\int_R \partial_t^* \varphi dt$$

is called Hitsuda-Skorokhod integral. We shall clarify the conditions on a non-random function f and on white noise functional φ.

Assume that φ is in $\mathcal{H}_n^{(-n)}$. Then,

$$\varphi = \langle F_n, :\dot{B}^{\otimes n}:\rangle,\ F_n \in \widehat{K^{-\frac{n+1}{2}}(R^n)}.$$

Proposition 5.2 *Suppose $F_n \otimes f$ is in $\widehat{K^{(-\frac{n+2}{2})}(R^{n+1})}$, then $\int f(t)\partial_t^*\varphi dt$ is defined and the integral is in $\mathcal{H}_n^{(-n+1)}$.*

Proof. Apply the S-transform to $\varphi(x)$ and $S(\partial_t^*\varphi)(x)$, we have

$$(S\varphi)(\xi) = \int \cdots \int_{R^n} F_n(u_1,\cdots,u_n)\xi(u_1)\cdots\xi(u_n)du^n$$

and the S-transform of $\partial_t^*\varphi$ is given by

$$S(\partial_t^*\varphi)(\xi) = \int \cdots \int_{R^{n+1}} F_n(u_1,\cdots,u_n)\delta_t(u_{n+1})\xi(u_1)\cdots\xi(u_{n+1})du^{n+1},$$

respectively.

Thus, we have

$$\partial_t^* \varphi = \langle F_n \otimes \delta_t, : \dot{B}^{\otimes(n+1)} : \rangle.$$

Then, we have the following correspondence under the S-transform.

$$\int f(t) \partial_t^* \varphi dt \longleftrightarrow F_n \otimes f.$$

Formally writing the integral of the form

$$\int_R f(t) \partial_t^* \varphi dt,$$

is called *Hitsuda-Skorokhod integral.*

■

We are going to show the advantage of introducing the integral in this form. It is a fact that the basic idea is to use the creation operator ∂_t^* which acts on the space $(L^2)^-$.

By using the creation operators, we can define Hitsuda-Skorokhod integrals

$$\int_R f(u) \partial_u^* \varphi(u) du,$$

where φ depends on u and not necessary to be non-anticipating. However, integrability requires some conditions on φ.

We restrict the t-variable to be in the unit interval [0,1] and will deal with Hitsuda-Skorokhod integral

$$\int_0^1 \partial_u^* \varphi(u) du.$$

See Hitsuda[61].

Theorem 5.4 *(Kubo-Takenaka) Suppose $\varphi(t)$ is non-anticipating and $E[\int_0^1 \varphi(t)^2 dt] < \infty$. Then*

$$\int_0^1 \varphi(t) \dot{B}(t) dt = \int_0^1 \partial_t^* \varphi(t) dt.$$

Example 5.4 $\int_0^1 \partial_t^* B(1)dt = B(1)^2 - 1.$

To prove the above expression, we first note that

$$\partial_t^* \dot{B}(s) =: \dot{B}(t)\dot{B}(s) :.$$

We therefore prove

$$\begin{aligned}\int_0^1 \partial_t^* B(1)dt &= \int_0^1 \partial_t^* \int_0^1 \dot{B}(s)dsdt = \int_0^1 \int_0^1 : \dot{B}(t)\dot{B}(s) : dsdt \\ &= \int_0^1 \int_0^1 (\dot{B}(t)\dot{B}(s) - \delta(t-s))dsdt \\ &= B(1)^2 - 1.\end{aligned}$$

We now extend Hitsuda-Skorokhod integral to the stochastic integral

$$\int_0^1 \partial_t^* \Phi(t)dt, \tag{5.4.1}$$

for $\Phi(t) \in (L^2)^-$.

The U-functional of (5.4.1) is obtained as

$$U\left[\int_0^1 \partial_t^* \Phi(t)dt\right](\xi) = \int_0^1 \xi(t)U[\Phi(t)](\xi)dt$$

which is well defined. Thus, we have defined the integral (5.4.1) as Hida distribution.

The following example is viewed as a generalization of the stochastic integral with respect to $\dot{B}(u)du$.

Example 5.5 If $\Phi(t) = \int_0^1 f(t,u) : \dot{B}(u)^2 : du$, then

$$\int_0^1 \partial_t^* \Phi(t)dt = \int_0^1 \int_0^t f(t,u) : \dot{B}(u)^2 \dot{B}(t) : dudt,$$

which can easily be extended to the integral with respect to $: \dot{B}(u)^n : du^n$.

An interesting application of the integral $\int_0^1 \partial_t^* \Phi(t)dt$ for $\Phi(t) \in (S)^*$ is the generalization of Itô's formula as is seen in the following theorem.

The Brownian motion $B(t)$ can be expressed as $B(t) = \langle x, 1_{[0,t]} \rangle, x \in S'(R), S'(R)$ being the dual space of the Schwartz space.

Theorem 5.5 *$F(B(t)) \in (S)^*$ for any $t \neq 0$ and $F \in S(R)$. Moreover, for any $0 < a \leq t$,*

$$F(B(t)) - F(B(a)) = \int_a^t \partial_s^* F'(B(s))ds + \frac{1}{2} \int_a^t F''(B(s))ds, \qquad (5.4.2)$$

where the derivatives are in the ordinary sense.

Proof. This assertion is a rephrasement of example 1 in Section 66 of Ito[65].

We may understand this fact by the computation in our terminology. Namely, setting $t = a + dt, dt > 0$,

$$\begin{aligned} dF(B(t)) &= F(B(a + dt)) - F(B(a)) = F(B(a) + \dot{B}(a)dt) \\ &= F'(B(a))\dot{B}(a)dt + \frac{1}{2}F''(B(a))\dot{B}(a)^2(dt)^2. \end{aligned}$$

Noting $\dot{B}(a)^2 =: \dot{B}(a)^2 : + \frac{1}{dt}$ and ignoring $: \dot{B}(a)^2 : (dt)^2$, we obtain

$$dF(B(t)) = \partial_t^* F'(B(t))dt + \frac{1}{2}F''(B(st))dt$$

which is equivalent to (5.4.2).

■

5.5 Lévy's stochastic integral

There is a stochastic integral which is defined by a somewhat different manner from what we have discussed so far.

It is a stochastic version of the Stieltjes integral. The integral due to P. Lévy [83] is defined as follows.

Let $u = f(x), v = g(x)$ be real valued function defined on an interval $[a, b]$, and let $X_1, X_2, \cdots, X_n, \cdots$ be a sequence of i.i.d. random variables which are all uniformly distributed over the interval $[a, b]$. Observed the values of $X_j, 1 \leq j \leq n-1$, and let them be arranged in an increasing order. The ordered statistics are $x_0^{(n)} = a, x_1^{(n)}, \cdots, x_{n-1}^{(n)}, x_n^{(n)} = b$.

Define

$$S_n = \sum_1^n \frac{1}{2}(u_{j-1} + u_j)(v_j - v_{j-1}),$$

where $u_j = f(x_j^{(n)})$ and $v_j = g(x_j^{(n)})$.

Set

$$S' = \liminf_{n\to\infty} S_n,$$

$$S'' = \limsup_{n\to\infty} S_n.$$

Then, there are non-random numbers I' and I'' such that

$$P(S' = I', S'' = I'') = 1.$$

The difference $I'' - I'$ is called the limit of the oscillation, denoted by $\lim_n \mathrm{Osc}(S_n)$.

Definition 5.1 If $I'' = I'$, then we say that udv is integrable in the stochastic sense (with respect to x), and I, the common value I' and I'', is called the stochastic integral of $f(x)dg(x)$ writing as

$$I = \int_a^b f(x)dg(x). \tag{5.5.1}$$

It is claimed that if both $f(x)$ and $g(x)$ are continuous, then the stochastic integral (5.5.1) exists, although for some cases the ordinary Stieltjes integral of this type does not exist.

It is interesting to compare P. Lévy's stochastic integral with the definition of the stochastic area discussed in Section 3.5.

5.6 Addendum : Path integrals

This section is devoted to the Feynman path integrals. They are not stochastic integral, but integration is done with respect to the white noise measure. In other words we discuss certain Hida distributions, which have quantum mechanical meanings, and their expectations. So, for the integrand, we meet normal quadratic functionals.

The trajectory of particle in classical mechanics is determined by the Lagrangian. If we come to quantum mechanics, we understand that there are many possible trajectories which are fluctuating around the classical trajectory. Following Dirac and Feynman, we understand that the quantum mechanical propagator is obtained by the average of $\exp(\frac{i}{\hbar}\int L(x)dt)$ over the possible trajectories (or paths) x.

Following this idea, the average (or integral) over the trajectories is considered within the framework of white noise theory. Fix the time interval to $[0, t]$. As is explained in the monograph[56] Section 10.1, the fluctuation around the classical trajectory is a Brownian bridge and the possible trajectories y are expressed in the form

$$y(s) = x(s) + \sqrt{\frac{\hbar}{m}}B(s),\ 0 < s < t, \tag{5.6.1}$$

where $B(s)$ is a Brownian motion, together with the pinning effect $y(t) = x(t)$ given by $\delta_0(B(t))$.

We now come to a quantum dynamics governed by the Lagrangian $L(x, \dot{x})$ which is of the form

$$L(x, \dot{x}) = \frac{1}{2}m\dot{x}^2 - V(x),$$

where $V(x)$ is potential.

The action $S[x]$ is evaluated on the time interval $[0, t]$:

$$S[x] = \int_0^t (\frac{1}{2}m\dot{x}(t)^2 - V(x(t)))dt.$$

Having been motivated by P. A. M. Dirac and R. Feynman, T. Hida and L. Streit proposed that quantum mechanical trajectory of $x(s)$ is expressed as a sum of classical trajectory and the Brownian bridge. Note that a Brownian bridge is a simple Markov Gaussian process. Namely, we take $y(s), 0 \le s \le t$ by (5.6.1), where $x(s)$ is the classical trajectory. We use the delta function $\delta_0(B(t))$ for the pinning effect of $B(s)$ to have a Brownian bridge.

The following example deals with normal ordinary quadratic functionals. Now we can see that a normal generalized functional naturally arises in mechanics.

Example 5.6 Consider a particle running with a constant velocity v in an environment interfered by a fluctuation expressed as $c\dot{B}(t)$. The velocity is therefore given by

$$\frac{1}{2}m(v + c\dot{B}(t))^2,$$

where m stands for the mass.

The total kinetic energy over the time interval $[0, T]$ is

$$\frac{1}{2}m\int_0^T (v + c\dot{B}(t))^2 dt.$$

But this quantity should be renormalized, so that we have a normal S-transform of the form

$$S\left(\frac{1}{2}m\left\{v^2T + 2vc\int_0^T \dot{B}(t)dt + c^2\int_0^T :\dot{B}(t)^2: dt\right\}\right)$$
$$= \frac{1}{2}m\left(v^2T + 2vc\int_0^T \xi(t)dt + c^2\int_0^T \xi(t)^2 dt\right).$$

This is an example of normal functional. This generalized quadratic functional is applied to the white noise approach to the Feynman path integral (see Streit and Hida[128]).

Theorem 5.6 *The quantum mechanical propagator $G = G(y_1, y_2; t)$ is expressed in the form*

$$G = E\left(Ne^{\frac{im}{2\hbar}\int_0^t x(s)^2 ds + \frac{1}{2}\int_0^t \dot{B}(s)^2 ds - \frac{i}{\hbar}\int_0^t V(x,s)ds}\,\delta(x(t) - y_2)\right), \quad (5.6.2)$$

where N is the factor of having the multiplicative renormalization and where $\frac{1}{2}\dot{B}(s)ds$ is necessary to have the white noise measure flatten.

Note. In the above expression (5.6.2), generalized functions are involved. So that, test functional

$$e^{-\frac{i}{\hbar}\int_0^t V(x,s)ds}$$

is put there.

Example 5.7 The harmonic oscillator

The potential is

$$V(x) = \frac{1}{2}gx^2.$$

The propagator is computed to have

$$G(0, y_2, t) = \left(\frac{m}{2\pi\hbar \sin \omega t}\right)\frac{1}{2} \ \exp\left[\frac{im\omega}{2\hbar}y_2^2 ctg\omega t\right]. \tag{5.6.3}$$

Chapter 6

Gaussian and Poisson Noises

A noise is a system of independent idealized elemental random variables parametrized by an ordered set. In this chapter we take the two typical noises, Gaussian and Poisson noises depending on the time parameter and we compare the construction of the measures corresponding to them. The construction itself illustrates the significant properties of these noises.

There is another intensity noise depending on the space parameter. It will be discussed in Chapter 8.

We first start with the construction of Poisson noise and then come to the Poisson measure which is characterized by symmetric group. In the course of its construction, quadratic forms of Gaussian variables play a basic role, which comes from a particular and in fact, interesting properties of Gaussian variables.

As a continuation we shall discuss duality of the spaces of quadratic forms of Gaussian noise.

6.1 Poisson noise and its probability distribution

We will apply the Bochner-Minlos Theorem to construct a Poisson noise measure space in the following.

Let E_1 and E_2 be suitably chosen Hilbert spaces ($\subset E_0 = L^2(R^d)$) that are topologized by Hilbertian norms $\|\cdot\|_1$ and $\|\cdot\|_2$, where

$$\|\xi\|_n^2 = \sum_{k=o}^{n} \int_{-\infty}^{\infty} (1+u^2)^n |\xi^{(k)}(u)|^2 du,$$

for $n = 0, 1, 2$.

Note that $\|\cdot\|_1 \le \|\cdot\|_2$ and that there exist consistent injections $T_i, i = 1, 2$, such that T_i from E_i to E_{i-1} for $i = 1, 2$, is of Hilbert-Schmidt type. Then, we have inclusions of the form

$$E_2 \subset E_1 \subset E_0 \subset E_1^* \subset E_2^*. \tag{6.1.1}$$

The characteristic functional of Poisson noise with R^d parameter with intensity λ is

$$C_P^d(\xi) = \exp\left[\lambda \int_{R^d} (e^{i\xi(t)} - 1)dt\right].$$

We can prove that $C_P^d(\xi)$ is continuous on E_1. Obviously it is positive definite and $C_P(0) = 1$. Thus by the Bochner-Minlos theorem, there exists a probability measure μ_P which is supported by E_2^* and then we have a Poisson noise measure space (E_2^*, μ_P).

To fix the idea, consider one dimensional parameter case and let the time parameter space be a compact set, say $I = [0, 1]$. In this case, $C_P^I(\xi)$ is continuous in $E_0 = L^2(I)$, so that there is a Gel'fand triple of the form

$$E_1 \subset L^2(I) \equiv E_0 \subset E_1^*. \tag{6.1.2}$$

A Poisson measure is now introduced on the space E_1^*.

Define $P(t, x) = \langle x, \chi_{[0,t]}\rangle, 0 \le t \le 1,\ x \in E_1^*$, by a stochastic bilinear form, where χ is the indicator function. Then, $P(t, x)$ is a Poisson process with parameter set $[0, 1]$.

Let A_n be the event on which there are n jump points over the time interval I. That is

$$A_n = \{x \in E_1^*; P(1, x) = n\}, \tag{6.1.3}$$

where n is any non-negative integer.

Then, the collection $\{A_n, n \ge 0\}$ is a partition of the entire space E_1^*. Namely, up to measure 0, the following relations hold:

$$A_n \bigcap A_m = \phi,\ n \ne m; \quad \bigcup A_n = E_1^*. \tag{6.1.4}$$

(More precisely, the above equalities hold modulo P.)

Given A_n, the conditional probability μ_P^n is defined since $\mu_P(A_n) > 0$, we have

$$\mu_P^n(A) = \frac{\mu_P(A_n \cap A)}{\mu_P(A_n)}, \ A \subset E_1^*.$$

For $C \subset A_k$, the probability measure μ_P^k on a probability measure space $(A_k, \mathcal{B}_k, \mu_P^k)$, is such that

$$\mu_P^k(C) = \mu_P(C|A_k) = \frac{\mu_P(C)}{\mu_P(A_k)},$$

where $\mathcal{B}_k$ is the sigma field generated by measurable subsets of A_k, determined by $P(t, x)$.

Let $x \in A_n, n \geq 1$, and let $\tau_i \equiv \tau_i(x), i = 1, 2, ..., n$, be the order statistics of jump points of $P(t)$:

$$0 = \tau_0 < \tau_1 < \cdots < \tau_n < \tau_{n+1} = 1.$$

(The τ_k's are strictly increasing almost surely.) Set

$$X_i(x) = \tau_i(x) - \tau_{i-1}(x),$$

so that

$$\sum_1^{n+1} X_i = 1.$$

Proposition 6.1 *(Si Si[114]) On the space A_n, the (conditional) probability distribution of the random vector $(X_1, X_2, ..., X_{n+1})$ is uniform on the simplex*

$$\sum_{j=1}^{n+1} x_j = 1, \quad x_j \geq 0.$$

Proof. It is known that the time intervals holding a constant value of a Poisson process $P(t), t \geq 0$, are all subject to an exponential distribution with density $\exp[-\lambda t], \lambda > 0, t \geq 0$, and they are independent. Hence, the joint distribution of holding times is subject to the direct product of exponential distribution.

$$P(X_1 \geq t_1, ..., X_{n+1} \geq t_{n+1}) = \prod \exp(-\lambda t_j) = \exp(-\lambda \sum_1^{n+1} t_j).$$

It can be easily seen that the distribution function of the joint distribution is constant over the simplex $\sum_1^{n+1} t_j = 1$. Thus the assertion is proved.

■

Corollary 6.1 *The probability distribution function of each X_j is*

$$1 - (1-x)^n, \quad 0 \le u \le 1.$$

Proposition 6.2 *The conditional characteristic functional*

$$C_{P,n}(\xi) = E[e^{i\langle \dot{P}, \xi \rangle} | A_n] \tag{6.1.5}$$

is obtained as

$$C_{P,n}(\xi) = \left(\int_0^1 e^{i\xi(t)} dt \right)^n. \tag{6.1.6}$$

Proof. By the definition of τ_j, we have

$$C_{P,n}(\xi) = E\left[e^{i \sum_1^n \xi(\tau_j)} \right],$$

while Proposition 6.1 gives us that the $e^{i\sum_1^n \xi(\tau_j)}$ has the same probability distribution as that of $e^{i\sum_1^n Y_j}$ by ignoring the order of τ_j's. Thus we have

$$E\left[e^{i \sum_1^n \xi(\tau_j)} \right] = E\left[e^{i \sum_1^n Y_j} \right].$$

Noting that Y_j's are independent and uniformly distributed, it can be proved that

$$C_{p,n}(\xi) = \left(\int_0^1 e^{i\xi(t)} dt \right)^n.$$

■

Construction of Poisson noise

Let a collection $\{A'_n, n \ge 0\}$ be any partition of the entire sample space Ω such that $P(A') = \frac{\lambda^n}{n!} e^{-\lambda}$.

Let $Y_k,\ 1 \le k \le n$, be a sequence of independent identically distributed random variables, on the probability space $(A'_n, P(\cdot|A'_n))$, which are uniformly distributed on $[0,1]$. The order statistics gives us an ordered sequence $Y_0 \le Y_{\pi(1)} \le Y_{\pi(2)} \le \cdots \le Y_{\pi(n)} \le Y_{\pi(n+1)}$, where $Y_0 = 0$ and

$Y_{\pi(n+1)} = 1$. Then, the probability distribution of $Y_{\pi(k+1)} - Y_{\pi(k)}$ has the distribution function $1 - (1 - x)^n$, for every $0 \le k \le n$. It can be proved by mathematical induction.

Theorem 6.1 *The process $\{Y_{\pi(k+1)} - Y_{\pi(k)},\ 1 \le k \le n,\}$ has the same distribution as that of $\{X_k,\ 1 \le k \le n\}$.*

This theorem gives us a way to construct a Poisson noise on I. In other words, with these Y_k's we can form a Poisson process $V^{(n)}(t, \omega)$, over the time interval [0,1], the kth jump point of which is $Y_{\pi(k)}$ in such a way that

$$V^{(n)}(t, \omega) = \sum_{k=1}^{n} \delta_{Y_{k(\omega)}}^{(n)}(t), \ \omega \in A_n. \tag{6.1.7}$$

Theorem 6.2 *Let a system $\{Y_j^{(n)}(\omega)\}$ be independent uniformly distributed random variables on I with $\omega \in A_N$. By arranging the $\{Y_j^{(n)}(\omega)\}$ in the order of increasing values, we have $V^{(n)}(t, \omega)$, by (6.1.7). Set*

$$V(t, \omega) = V^{(n)}(t, \omega), \quad \omega \in A_n'; \ n = 0, 1, ... \tag{6.1.8}$$

on I. Then, $V(t, \omega), \omega \in \Omega$, is a Poisson noise with the parameter t running through I.

6.2 Comparison between the Gaussian white noise and the Poisson noise, with the help of characterization of measures

It is known that Gaussian noise (white noise) can be characterized by the infinite dimensional rotation group from the viewpoint of infinite dimensional harmonic analysis. In reality, the group plays essential roles in white noise analysis. While, we have seen (in the literature[114]) that symmetric groups are characteristics of Poisson noise. We first briefly recall the construction of Gaussian white noise measure and Poisson noise measure in what follows for comparison.

(1) Gaussian white noise

Gaussian white noise can be constructed on the projective limit space of spheres. We start with the probability measure space (S^1, m_1), where S^1

is the unit circle and m_1 is $\frac{d\theta}{2\pi}$. We construct a probability measure space (S^n, m_n) by projection where S^n is an n-dimensional sphere without containing north and south poles, and m_n be the uniform probability measure on S^n.

Take the projection π_n from S^{n+1} down to S^n is defined such as each longitude is projected to the intersection point of the longitude and the equator which can be viewed as the equator.

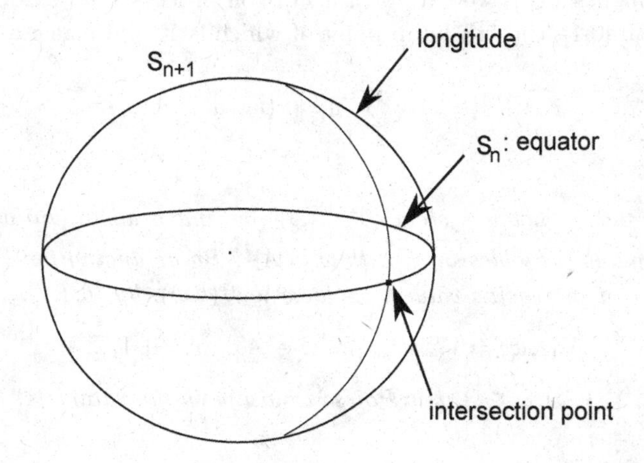

Fig. 6.1

Suppose the measure m_n is given. Then, m_{n+1} on S^{n+1} is determined in such a way that

$$m_{n+1}(\pi_n^{-1}(E)) = m_n(E), E \in S_n.$$

It can also be said that m_n is the marginal distribution of m_{n+1} and with this restriction m_{n+1} has the maximum entropy; that is, the uniform distribution, which is invariant under the rotation group $SO(n+1)$.

Thus, the consistent family $\{(S_n, m_n, \pi_n)\}$ defines the projective limit space $\{(S_\infty, m)\}$. The limit can be identified with the white noise measure space (E^*, μ), with the characteristic functional

$$C(\xi) = e^{-\frac{1}{2}\|\xi\|^2},$$

where E^* is the space of generalized functions.

(2) Poisson noise

Consider a Poisson noise $\dot{P}(t), t \in (0,1]$. We have shown in the previous section that the joint probability distribution of $(X_1, X_2, \cdots, X_{n+1})$ is the uniform distribution over the n-dimensional simplex

$$\Delta_n = \{(x_1, \cdots, x_{n+1}); \sum_{i=1}^{n+1} x_i = 1, x_i > 0\}. \tag{6.2.1}$$

Uniform probability distribution on a simplex and the invariance of the conditional Poisson distribution under the symmetric group are significant to characterize a Poisson noise. With this fact we have constructed a Poisson noise $\dot{P}(t), t \in (0,1]$ in such a way that we decompose a measure space defining the noise and on each component of the decomposition we can see those characteristics in a visualized manner.

Start with $n = 1$. A jump point is distributed uniformly on [0,1] so that a random jump point has maximum entropy. That is we have a probability space (Δ_1, μ_1), $\Delta_1 = \{(x_1, x_2); x_i \geq 0,\ x_1 + x_2 = 1\}$ and μ_1 is the Lebesgue measure.

Let μ_n be the joint distribution of $(X_1, \cdots, X_{n+1})$ on the simplex Δ_n. Then we can determine the joint distribution of $X_1, \cdots, X_{n+1}$ having maximum entropy, that is uniform distribution.

Let π_n be the projection of Δ_{n+1} down to Δ_n which is a side simplex of Δ_{n+1}, determined as follows. Given a side simplex Δ_n of Δ_{n+1}. Then, there is a vertex v_n of Δ_{n+1} which is outside of Δ_n. The projection π_n is a mapping defined by

$$\pi_n : \overline{v_n x} \to x,$$

where $\overline{v_n x}$ is defined as a join connecting the v_n and a point x in Δ_n.

Let μ_{n+1} be a probability measure such that

$$\mu_{n+1}(\pi_n^{-1}(B)) = \mu_n(B), \quad B \subset \Delta_n$$

and that μ_{n+1} has maximum entropy under the boundary condition.

Since there is a freedom to choose a boundary of Δ_{n+1}, the requirements on μ_{n+1}, the measure space $(\Delta_{n+1}, \mu_{n+1})$ should be invariant under symmetric group $S(n+2)$ which acts on permutations of coordinates x_i.

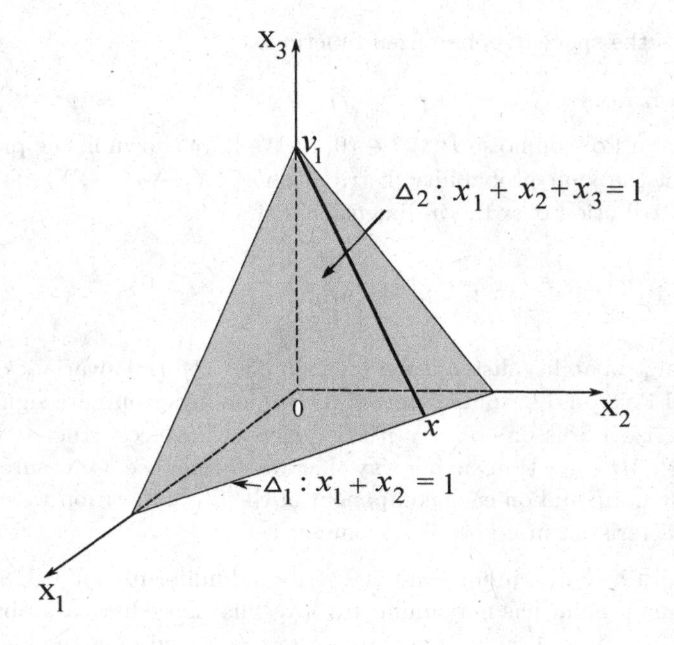

Fig. 6.2

In this way we can form $\{(\Delta_n, \mu_n)\}$ successively by using the projection $\{\pi_n\}$, where μ_n is to be the uniform probability measure on Δ_n.

Theorem 6.3

i) The measure space (Δ_n, μ_n) is isomorphic to the measure space $(A'_n, P(\cdot|A'_n))$, defined in the previous section.

ii) The weighted sum (Δ_∞, μ) of measure spaces $(\Delta_n, \mu_n), n = 1, 2, \cdots$ is identified with the Poisson noise space.

This method can be compared to the approximation to Gaussian white noise by projective limit of spheres, where one can see the characteristics; rotation invariant and maximum entropy. The comparison is successful to observe the two noises step by step. It can be seen in the following table.

Gaussian	Poisson
Start with$(S_1, d\theta)$, S_1 : circle, $d\theta = \frac{1}{2\pi}$.	Start with(Δ_1, μ_1), $\Delta_1 = \{(x_1, x_2); x_1 + x_2 = 1\}$
$(\tilde{S}_n, m_n)$, $\tilde{S}_n$: n-dimensional sphere, excluding north and south poles dm_n : uniform prob measure on $\tilde{S}_n$.	(Δ_n, μ_n), Δ_n : n-simplex $\Delta_n \subset R^{n+1}$, μ_n : uniform distribution on Δ_n
projection $\pi_n : \tilde{S}_{n+1} \mapsto \tilde{S}_n$	projection $\pi_n : \Delta_{n+1} \mapsto \Delta_n$
$m_{n+1}(\pi_n^{-1}(E)) = m_n(E), E \subset \tilde{S}_n$	$\mu_{n+1}(\pi_n^{-1}(B)) = \mu_n(B), \quad B \subset \Delta_n$
m_{n+1} : uniform distribution on $\tilde{S}_{n+1}$	μ_{n+1} : uniform distribution on Δ_n
$SO(n+2)$ invariant	$S(n+2)$ invariant
Form $(\tilde{S}_n, m_n)$ successively by using the projection $\{\pi_n\}$	Form $\{(\Delta_n, \mu_n)\}$ successively by using the projection $\{\pi_n\}$
The limit $(\tilde{S}_n, m_n)$ is (E^*, μ) Characteristic functional : $e^{-\frac{1}{2}\|\xi\|^2}$ $\Rightarrow$ Existence of Gaussian measure μ	$\Delta_\infty = \cup_n \Delta_n$ $\mathcal{B}(\Delta_\infty) = \sigma\{\cup A_n,\ A_n \in \mathcal{B}(\Delta_n)\}$ $\mu(A) = \sum_n p_n \mu_n(A \cap \Delta_n),\ A \in \mathcal{B}(\Delta_\infty)$ where $p_n = \frac{\lambda^n}{n!}, \quad \lambda > 0$
(E^*, μ) : Space of generalized functions	(Δ_∞, μ): Poisson noise space

6.3 Symmetric group in Poisson noise analysis

We recall our results in the paper[116] which are important background for the discussion related to the symmetric group.

When the characterization of Poisson noise is discussed in the previous section, the infinite symmetric group $S(\infty)$ plays an important role. Thus we understood $S(\infty)$, the limit of $S(n)$, with the help of its unitary representation. We will briefly describe this fact in the following.

Let a unitary representation of $S(n)$ be given on R^n in such a way that for each $\pi \in S(n)$, define $U_\pi^{(n)}$ such that

$$U_\pi^{(n)}(\sum_{k=1}^{n} x_k e_k) = \sum_{k=1}^{n} x_k e_{\pi(k)}, \tag{6.3.1}$$

where e_k's form a base of R^n.

It can easily be seen that $U_\pi^{(n)}$ is a unitary representation of π on R^n. To form a simplest irreducible factor of the representation we take a subset R_1^n of R^n such that

$$R_1^n = \{c \sum_1^n e_i, c \in R\}.$$

Then, we are given $\widetilde{U}_\pi^{(n)}$, on R_1^n, such that

$$\widetilde{U}_\pi^{(n)}(c \sum_{k=1}^{n} e_k) = c \sum_{k=1}^{n} e_{\pi(k)}, \tag{6.3.2}$$

for each $\pi \in S(n)$.

We can see that $\widetilde{U}_\pi^{(n)}$ defines an irreducible unitary representation of $S(n)$ on R_1^n.

Take a family of pairs $\{(S(n), \widetilde{U}^{(n)})\}$. We then follow Bochner's method to obtain the projective limit of $\{(S(n), \widetilde{U}^{(n)})\}$. For this purpose we have to find a base $\{e_k\}$ for which we can determine a family of projections $\{f_{m,n}\}$ such that

$$f_{mn} : (S(n), \widetilde{U}^{(n)}) \longmapsto (S(m), \widetilde{U}^{(m)}),$$

where $m < n; m, n \in \mathbf{N} = \{1, 2, \cdots\}$.

There is no restriction to choose the base $\{e_n\}$, however we will choose such a base that is formed by quadratic Hida distributions, $\{: \Delta_k^{(n)} X_n(t)^2 :\}$ for $\{e_n\}$ in (6.3.2), where $X_n(t)$ is Gaussian which approximates Brownian motion $B(t)$ in Lévy's construction of Brownian motion (see Chapter 2).

The reader might think somewhat strange to take Gaussian quadratic form in the discussion of Poisson noise. To be surprised the choice of the base gives a big advantage that comes from the following fundamental lemma for the square of Gaussian random variables.

Lemma 6.1 *Let X and Y be mutually independent random variables that are subject to the standard Gaussian distribution. Then we have*

$$E(X^2 + Y^2 - 2|(X+Y)^2 - 2) = \frac{1}{2}(X+Y)^2 + 1.$$

Proof. It is sufficient to take the projection of $X^2 + Y^2$ down to the space spanned by $(X+Y)^2 - 2$. It can be done within $\mathcal{H}_2$. We can see, by computation, that

$$X^2 + Y^2 - 2 - \frac{1}{2}\{(X+Y)^2 + 2\} \tag{6.3.3}$$

is orthogonal to in fact, independent of the space generated by the variable $(X+Y)^2 - 2$. This property proves the lemma. ■

Proposition 6.3 *The following relation holds.*

$$E\left(:\Delta_{2k}^{(n+1)} X_{n+1}(t)^2 : + : \Delta_{2k+1}^{(n+1)} X_{n+1}(t)^2 : \,\middle|\, \mathcal{B}_n(Y) \right) = \frac{1}{2} : \Delta_k^{(n)} X_n(t)^2 : . \tag{6.3.4}$$

Proof comes from Lemma 6.1. Note that the formula (6.3.3) is useful to prove equation (6.3.4).

We are now ready to define the projection. Define

$$f_{n+1,n} : (S(2^{n+1}), \widetilde{U}^{(2^{n+1})}) \mapsto (S(2^n), \widetilde{U}^{(2^n)})$$

in such a way that for $\pi^{(2^{n+1})} \in S(2^n)$,

$$f_{n+1,n} : \pi^{(2^{n+1})} \;\mapsto\; \pi^{(2^n)} \in S(2^n),$$

where $\pi^{(2^n)}$ is obtained by deleting all even numbers j in $\pi^{(2^{n+1})}(j)$. Hence only those members such that $\pi^{(2^{(n+1)})}(2j+1) = 2k+1$ remain. We therefore have

$$f_{n+1,n}\left(\widetilde{U}^{(2^{n+1})}\right) = \widetilde{U}^{(2^n)}$$

and we obtain $\widetilde{U}^{(2^n)}$ by using the conditional expectation

$$E(\,\cdot \mid \mathcal{B}_{2^n}(Y)).$$

With the special choice of the base which is characterized by Proposition 6.3, this mapping $f_{n+1,n}$ defines a projection.

Define $f_{n,m}, n > m$ such that

$$f_{n,m} = f_{n,n-1}f_{n-1,n-2}\cdots f_{m+1,m}.$$

Then, the collection $\{f_{n,m}, n > m\}$ is a consistent family of projections. Hence we can prove the following theorem.

Theorem 6.4 *The sequence $\{(S(2^n), \widetilde{U}^{2^n})\}$ defines a projective limit, which is denoted by $(S(\infty), \widetilde{U}^{\infty})$.*

Remark 6.1 *The choice of the family of : $(\Delta_k X(t))^2$: (i.e. renormalized square of $\Delta_k X(t)$) enables us to define a projection by the particular property, stated in Lemma 6.1, for Gaussian random variables.*

Proposition 6.4 (*Si Si*[114]) *The conditional probability measure μ_P^n, defined on the measure space $(A_n, \mathbf{B}_n, \mu_P^n)$, is invariant under the symmetric group $S(n+1)$ acting on $(X_1, X_2, ..., X_{n+1})$.*

6.4 Spaces of quadratic Hida distributions and their dualities

This section returned to the Gaussian case to consider the duality within the space of quadratic Hida distributions. Again we take $I = [0, 1]$.

Let us remind the Gel'fand triple (See [8] Chapter 2) to form

$$\mathcal{H}_2^{(2)} \subset \mathcal{H}_2 \subset \mathcal{H}_2^{(-2)},$$

where

$$\mathcal{H}_2^{(2)} = \left\{ \int\int_{I^2} F(u,v) : \dot{B}(u)\dot{B}(v) : dudv, F \in \widehat{K^{\frac{3}{2}}(I^2)} \right\},$$

$$\mathcal{H}_2 = \left\{ \int\int_{I^2} F(u,v) : \dot{B}(u)\dot{B}(v) : dudv, F \in \widehat{K^2(I^2)} \right\},$$

and

$$\mathcal{H}_2^{(-2)} = \left\{ \int\int_{I^2} F(u,v) : \dot{B}(u)\dot{B}(v) : dudv, F \in \widehat{K^{-\frac{3}{2}}(I^2)} \right\},$$

in which $\widehat{\cdot}$ means symmetric.

We use the notation $\widehat{K^m(I^2)}$ to express the Sobolev space of order m over I involving symmetric function, where m can be positive or negative rational number.

Define a new subspace $\mathcal{H}_2^{(-2,1)}$ of $\mathcal{H}_2^{(-2)}$:

$$\mathcal{H}_2^{(-2,1)} = \left\{ \int_I f(u) : \dot{B}(u)^2 : du, f \in L^2(I) \right\}.$$

The function f may be viewed as $f(\frac{u+v}{2})\delta(u-v) \equiv \widetilde{f}(u,v)$ to claim that the quadratic form is in $\mathcal{H}_2^{(-2)}$.

It is known that (see the monograph[57]) the following equality holds :

$$\langle \int_I f(u) : \dot{B}(u)^2 : du, \int_I g(u) : \dot{B}(u)^2 : du \rangle_{\mathcal{H}_2^{(-2)}} = 2(\widetilde{f}, \widetilde{g})_{K^{-\frac{3}{2}}(I)}.$$

Dual space of $\mathcal{H}_2^{(-2,1)}$

As was noted before, we may replace $X_n(t)$ with $B(t)$ as far as increments over $\Delta^{(n)}$ are concerned. For convenience, we shall use $B(t)$ from now on.

Since (stochastic) integrals of the form

$$\int_I f(t) : \dot{B}(t)^2 : (dt)^2$$

are formal at present, so that an interpretation of them should be given. For this purpose we prepare some more background.

Set

$$\mathcal{H}_2^{(-2,2)} = \left\{ \int_I g(u) : \dot{B}(u)^2 : (du)^2, g \in L^2(I) \right\}.$$

By using Proposition 6.3 and noting

$$|\Delta_{2k}^{(n+1)}| = |\Delta_{2k+1}^{(n+1)}| = 2^{-(n+1)} (= |\Delta^{(n+1)}|),$$

we have the following conditional expectation.

$$E\left(: \left(\frac{\Delta_{2k}^{(n+1)} B}{\sqrt{\Delta^{(n+1)}}} \right)^2 : + : \left(\frac{\Delta_{2k+1}^{(n+1)} B}{\sqrt{\Delta^{(n+1)}}} \right)^2 : \Big| : \left(\frac{\Delta_k^{(n)} B}{\sqrt{\Delta^{(n)}}} \right)^2 : \right) =: \left(\frac{\Delta_k^{(n)} B}{\sqrt{\Delta^{(n)}}} \right)^2 :$$

We can therefore prove

$$E\left[\sum_k \left(: \left(\frac{\Delta_{2k}^{(n+1)} B}{\sqrt{\Delta^{(n+1)}}} \right)^2 : + : \left(\frac{\Delta_{2k+1}^{(n+1)} B}{\sqrt{\Delta^{(n+1)}}} \right)^2 : \left| \frac{:(\Delta_k^{(n)} B)^2:}{\Delta^{(n)}}, 0 \le k \le 2^n - 1 \right) \right]$$

$$= \sum_k E\left(: \left(\frac{\Delta_{2k}^{(n+1)} B}{\sqrt{\Delta^{(n+1)}}} \right)^2 : + : \left(\frac{\Delta_{2k+1}^{(n+1)} B}{\sqrt{\Delta^{(n+1)}}} \right)^2 : \left| \frac{:(\Delta_k^{(n)} B)^2:}{\Delta^{(n)}} \right. \right)$$

$$= \sum_k : \left(\frac{\Delta_k^{(n)} B}{\sqrt{\Delta^{(n)}}} \right)^2 : .$$

Hence

$$E\left(\sum_k : \left(\frac{\Delta_k^{(n+1)} B}{\Delta^{(n+1)}} \right)^2 : \Delta^{(n+1)} \right| : (\Delta_k^{(n)} B)^2 :, \ 0 \le k \le 2^n - 1 \Bigg)$$

$$= \sum_k : \left(\frac{\Delta_k^{(n)} B}{\Delta^{(n)}} \right)^2 : \Delta^{(n)}$$

Thus, for the quadratic form

$$\sum_j : \left(\frac{\Delta_j^{(n)} B}{\Delta^{(n)}} \right)^2 : \Delta^{(n)},$$

we prove that its average

$$\frac{1}{2^n} \sum_j : \left(\frac{\Delta_j^{(n)} B}{\Delta^{(n)}} \right)^2 : \Delta^{(n)} = \sum_j : \left(\frac{\Delta_j^{(n)} B}{\Delta^{(n)}} \right)^2 : (\Delta^{(n)})^2 \tag{6.4.1}$$

is consistent in n by the projection which is realized by the conditional expectation $E(\cdot|\mathbf{B}_n)$, where $\mathbf{B}_n$ is generated by $: (\Delta_j^{(n)} B)^2 :, 0 \le j \le 2^n - 1$.

Hence, we conclude that the projective limits exist and they are

$$\int_I : \dot{B}(t)^2 : dt \tag{6.4.2}$$

and

$$\int_I : \dot{B}(t)^2 : (dt)^2 \tag{6.4.3}$$

in the sense that (6.4.2) is a limit in $\mathcal{H}_2^{(-2)}$, while (6.4.3) exists only as a *projective limit* of (6.4.1). (The formula (6.4.3) is a formal expression.)

Now, it should be noted that these two are $S(\infty)$ *invariant.*

By the usual method of integration, we can define

$$X = \int_I f(t) : \dot{B}(t)^2 : dt. \tag{6.4.4}$$

Also, projective limit of (6.4.1) in the space $\mathcal{H}_2^{(-2)}$ defines

$$Y = \int_I g(t) : \dot{B}(t)^2 : (dt)^2, \tag{6.4.5}$$

provided that f and g are measurable.

Before letting n tend to ∞, at any stage n, we have approximations to (6.4.4) and (6.4.5) as follows.

$$\sum a_k : \left(\frac{\Delta^{(n)}B}{\Delta^{(n)}}\right)^2 : \Delta^{(n)} = I_n$$

$$\sum b_k : \left(\frac{\Delta^{(n)}B}{\Delta^{(n)}}\right)^2 : (\Delta^{(n)})^2 = J_n$$

$$\langle I_n, J_n \rangle \equiv E(I_n J_n) = \sum a_k \bar{b}_k E\left(: \left(\frac{\Delta^{(n)}B}{\Delta^{(n)}}\right)^2 : \right)^2 (\Delta^{(n)})^3$$

where $\langle\ ,\ \rangle$ is the bilinear form connecting $\mathcal{H}_2^{(-2,1)}$ and $\mathcal{H}_2^{(-2,2)}$. Use the evaluation

$$E\left(: \left(\frac{\Delta_k^{(n)}B}{\Delta^{(n)}}\right)^2 : \right)^2 = 2\frac{1}{(\Delta^{(n)})^2},$$

we have

$$\langle I_n, J_n \rangle = 2\sum a_k \bar{b}_k \Delta^{(n)} \to 2\int_I f(t)\overline{g(t)}dt,$$

as $n \to \infty$.

Here we approximate the functions f and g expressed in (6.4.4) and (6.4.5). This limit is independent of the choice of other approximations of f and g.

Thus, $\mathcal{H}_2^{(-2,2)}$ can be seen as the dual space of $\mathcal{H}_2^{(-2,1)}$.

Chapter 7

Multiple Markov Properties of Generalized Gaussian Processes and Generalizations

From the viewpoint of dependency, the Markov property comes after additivity or independent sequences. Then comes multiple Markov properties.

We restrict, in this chapter, our attention to Gaussian processes and discuss their multiple Markov properties which are typical properties to describe the way of dependency of Gaussian processes according to the passage of the time. For this purpose, we assume that the Gaussian processes in question have the canonical representation in terms of white noise.

There is a general theory of the canonical representation of a Gaussian process $X(t), t \in T$, where T is an interval, a subset of R^1, which is purely non-deterministic and separable (see Hida[21]). In order to establish good connections between multiple Markov properties and the canonical representation, we assume that $X(t)$ has unit multiplicity with continuous spectrum.

We can therefore assume that $X(t)$ is expressed in the form

$$X(t) = \int^{t} F(t,u)dZ(u),$$

where $Z(u)$ is an additive Gaussian process and that for any $t > s$,

$$E(X(t)|\mathcal{B}_s(X)) = \int^{s} F(t,u)dZ(u),$$

where $\mathcal{B}_s(X)$ is the σ-field generated by $X(u), u \leq s$.

To fix the idea we further assume that the canonical representation of $X(t)$ is expressed in the form, by taking $Z(u) = B(u)$, $B(u)$ being a

Brownian motion and hence

$$X(t) = \int^t F(t,u)dB(u)$$

by the Wiener integral, or equivalently

$$X(t) = \int^t F(t,u)\dot{B}(u)du,$$

where the domain of integration is $\{s \le t; s \in T\}$ and simply denoted by $\int^t$.

The canonical representation is fitting, as we have discussed before, for the idea of reduction which leads us to express the given process as a functional of independent random variables; the white noise $\dot{B}(t)$ in this case.

The way of dependence of a Gaussian process as time t goes by can well be expressed by the canonical representation. In particular, the Markov property as well as multiple Markov properties are the best properties where the canonical representation serves most efficiently.

The analytic expression of the canonical representation of a Gaussian process leads us to two directions; one is a more general case, namely generalized Gaussian processes, where we can find a dual process and the other is that the particular analytic form of the canonical kernel can be used to describe a dual process which has also multiple Markov property of the same degree.

In addition, we can see a typical, beautiful duality between time and space through the multiple Markov properties.

7.1 A brief discussion on canonical representation theory for Gaussian processes and multiple Markov property

Before we discuss generalized white noise functionals, we will recall the related topics on classical Gaussian processes.

I. Canonical representation theory for Gaussian processes

Here, we recall the canonical representation in line with reductionism. See the monograph[57].

First, we will explain the idea of the canonical representation theory for Gaussian processes.

The theory has been originated by P. Lévy, in his lecture at the third Berkeley Symposium on Mathematical Statistics and Probability, 1955 (Proceedings 1956). His motivation was the study of multidimensional parameter Brownian motion $X(A), A \in R^d$, call it Lévy Brownian motion. One of the tools of the study is to form the average of $X(A)$ over the sphere $S(t)$ of radius t with center at origin. The average is defined by the integral over $S(t)$ with uniform probability measure. The average is denoted by $M_d(t), t \geq 0$.

Lévy's idea of investigation of $M_d(t)$ is to express it as a canonical form, namely as a stochastic integral of the form

$$M_d(t) = \int_0^t F(t,u)\xi_u\sqrt{du}, \tag{7.1.1}$$

where $F(t, u)$ is a non-random function defined on $\{(t, u), u \leq t\}$, that is a Volterra kernel.

The integral based on the random measure $\xi_u\sqrt{du}$, which may be written as $dB(t)$, has been explained rigorously in Remark 5.1. For the expression

$$\int f(u)\xi_u\sqrt{du},$$

we can give interpretation in a similar manner to the Wiener integral and so it can be expressed as

$$\int f(u)dB(u).$$

We now back to equation (7.1.1). P. Lévy used to think of the canonical form of a Gaussian random variable X which is to be expressed as a sum

$$X = \mu + \sigma\xi,$$

where μ is the conditional expectation, ξ is a standard Gaussian random variable and σ is the standard deviation.

With this idea Lévy gave the requirement that the representation (7.1.1) should have the property, for $s < t$,

$$E(M_d(t)|\mathcal{B}_s(M_d)) = \int^s F(t,u)\xi\sqrt{du}. \tag{7.1.2}$$

The representation with this property is called the **canonical representation.**

Hida's idea on the study of Gaussian processes seems much influenced by Lévy's approach to the representation theory of Gaussian processes. In addition, he had the idea of obtaining an expression of a given Gaussian process $X(t)$ in terms of Brownian motion, which is quite familiar for us and a non-random kernel function $F(t,u)$ to have an integral

$$X(t) = \int^t F(t,u)dB(u).$$

Further he has recognized the significance of canonical representation, of course uniqueness is important and to be more significant, the increment dB turns out to be the *innovation* of $X(t)$ if the representation is canonical. Such a consideration has led to his guiding technique for the analysis of stochastic process. In terms of his recent expression it can be expressed as

Reduction → Synthesis → Analysis.

The notion of the innovation will be explained later.

Factorization of the covariance function is one of the main tools for the study of stationary Gaussian processes (see the monograph[57]).

The following example has given a suggestion to recognize the meaning of the representation by observing the two Gaussian processes due to Lévy.

Example 7.1

$$X_1(t) = \int_0^t (2t-u)dB(u), \tag{7.1.3}$$

$$X_2(t) = \int_0^t (-3t+4u)dB(u). \tag{7.1.4}$$

To be surprised, two processes are the same. In fact, the two processes have the same covariance $3ts^2 - \frac{2}{3}s^3$ for $t > s \geq 0$.

Most important viewpoint is that the representation (7.1.3) of $X_1(t)$ is *canonical*, but (7.1.4) for $X_2(t)$ is not.

There are many important cases which suggested to claim that $B(t)$ is a standard Brownian motion, so that $dB(u)$ may be written as $\dot{B}(u)du$, where $\dot{B}(u)$ is a white noise.

Definition 7.1 A representation of $X(t)$, given by

$$X(t) = \int^t F(t,u)\dot{B}(u)du, \tag{7.1.5}$$

is said to be *canonical* if the following equality for the conditional expectation holds for every $t > s$:

$$E\Big(X(t) \Bigm| \mathcal{B}_s(X)\Big) = \int^s F(t,u)\dot{B}(u)du. \tag{7.1.6}$$

A criterion for the canonical property on the kernel $F(t,u)$ is to be given in Theorem 7.1.

The canonical property of a representation was proposed by P. Lévy in 1955 at the third Berkeley Symposium on Math. Statistic and Probability. General theory of existence was given by T. Hida[21] in 1960 and later by H. Cramér in 1961. Useful applications of this theory are found in the theory of multiple Markov property and in the study of Lévy's Brownian motion. They are also given in the paper[88] and others.

The problem of obtaining the canonical kernel has close connection with the factorization of the covariance function; indeed getting the *optimal* kernel among the possible factorizations. One of the ideas of the factorization is the use of the theory of Reproducing Kernel Hilbert Space (RKHS) which is briefly discussed in Chapter 3.

To fix the idea, for ordinary Gaussian process $X(t)$, we assume that it has multiplicity one.

Theorem 7.1 *Hida* (1960) **(Criterion of canonical property)**

A representation in the form (7.1.5) is a proper canonical representation if and only if, for any fixed $t_0 \in T$ *and* φ *in* $L^2(0,t_0)$

$$\int^t F(t,u)\varphi(u)du = 0 \quad \textit{for every } t \le t_0$$

implies

$$\varphi(u) = 0 \ a.e. \ on \ (-\infty, t_0]^T.$$

Proof is referred to Hida[21].

II. Simple Markov Gaussian processes

We first recall the simple Markov property of Gaussian process which can be given by the equivalent condition as follows.

A Gaussian process is simple Markov if

$$E[X(t)|\mathcal{B}_s(X)] = \varphi(t,s)X(s), \quad s \le t, \tag{7.1.7}$$

where $\varphi(t,s)$ is a real ordinary function of (t,s).

Here we can prove that

$$\varphi(t,s) = \frac{f(t)}{f(s)} \text{ for } t > s.$$

With some regularity assumptions, we can prove that a martingale $U(t)$ has a canonical representation of the form

$$U(t) = \int^t g(u)\dot{B}(u)du,$$

where g is locally square integrable. For a simple Markov process $X(t)$ we have

$$X(t) = f(t)U(t) = f(t)\int^t g(u)\dot{B}(u)du,$$

which is a canonical representation.

The canonical representation is obtained by Reproducing Kernel Hilbert Space technique. It is unique and is of the form (7.1.5). This means that a Gaussian process can be completely characterized by the kernel F and white noise. The former is investigated by non-random functional analysis. The latter is the most basic stochastic process (generalized process, indeed) and its probabilistic properties are well-known.

Hence the combination of the studies of F and referring the probabilistic properties of $\dot{B}$ we can investigate Gaussian process.

There is an interesting case where dependence (probabilistic property) determines analytic structure of the kernel.

III. Multiple Markov Gaussian processes

We come to the study of general multiple Markov Gaussian processes by using the canonical representation. In what follows, in this chapter, we consider the Gaussian processes which have canonical representation using a Brownian motion.

Definition 7.2 For any choice of $\{t_i\}$ and $t_0 \le t_1 < ... < t_N < t_{N+1}$,

i) $E[X(t_i)|\mathcal{B}_{t_0}(X)], i = 1, 2, ..., N$, are linearly independent and

ii) $E[X(t_i)|\mathcal{B}_{t_0}(X)], i = 1, 2, ..., N+1$, are linearly dependent

then, $X(t)$ is said to be *N-ple Markov* Gaussian process.

A Gaussian process satisfying the N-ple Markov property (multiple Markov property) can be expressed as a sum of additive processes $U_i(t)$ with the coefficients $f_i(t)$ which are sure functions :

$$X(t) = \sum_{i=1}^{N} f_i(t) U_i(t).$$

More precisely we can establish the theorem.

Theorem 7.2 *(Hida 1960) The process $X(t)$ is N-ple Markov if and only if the kernel F, in the expression (7.1.6) is*

$$F(t, u) = \sum_{1}^{N} f_i(t) g_i(u) \tag{7.1.8}$$

where $det(f_i(t_j)) \neq 0$ for different t_j's and $\{g_i(u), i = 1, 2, \cdots\}$ are linearly independent on $L^2([0, t])$, for any t.

For the proof we refer to the literature[21].

Definition 7.3 The kernel expressed in Theorem 7.2 is called a Goursat kernel of order N.

Thus, we claim that the way of the propagation of randomness through an N-ple Markov Gaussian process can be expressed with a Goursat kernel $\sum_1^N f_i(t)g_i(u)$.

In general, the two systems $\{f_i(t)\}$ and $\{g_i(u)\}$ show different properties. However, we can see a duality between the two systems for some special processes like stationary processes and strictly multiple Markov processes. Such a duality stands for interesting probabilistic properties namely, we can form a dual process.

The idea for the dual property is that the function $g_i(u)$ is the weight to collect the past independent random variables $\dot{B}(t)$ up to time t. On the other hand $f_i(t)$ is the weight for the future t. Their roles are different; one is for the past and the other is for the future.

IV. The N-ple Markov property in the restricted sense

We rephrase Doob's approach to multiple Markov property in terms of white noise.

A Gaussian process $X(t)$ is said to be **N-ple Markov in the restricted sense**, if $X(t)$ is differentiable in strong topology up to $(N-1)$ times and satisfy

$$L_t X(t) = \dot{B}(t)$$

for an N-th order differential operatorL_t. The N-th derivative is taken in the topology of Hida distribution.

Let $D = \frac{d}{dt}$ and L_t be an N-th order ordinary differential operator such that

$$L_t = \sum_{k=0}^{N} a_k(t) D^{N-k},$$

where a_k's are sufficiently smooth.

In our setup, the differential equation is well defined in $\mathcal{H}_1^{(-1)}$, Hida distribution space which is isomorphic to Sobolev space of order -1. For instance, $B(t)$ is differentiable in t in $\mathcal{H}_1^{(-1)}$ since $\chi_{[0,1]}(u)$ is differentiable in the Sobolev space and the derivative is $-\delta_t$.

Theorem 7.3 *There is a solution to the differential equation*

$$L_t X(t) = \dot{B}(t)$$

expressed in the form

$$X(t) = \int_0^t R(t,u)\dot{B}(u)du, \qquad (7.1.9)$$

where $R(t,u)$ is the Riemann function for an ordinary differential operator L_t. The representation of $X(t)$ is canonical.

Obviously the solution $X(t)$ satisfies the N-ple Markov property, so that it is said to be an N-ple Markov Gaussian process in the restricted sense.

The above theorem is used in Section 4.2 to discuss the stochastic differential equation. To establish the duality for Gaussian processes, we need to discuss the following computation in detail.

We can take real valued functions $v_i(t), i = 0, 1, 2, \cdots, N$, such that (see Hida[21])

$$v_i(t) \in \boldsymbol{C}^N, \ i = 0, 1, 2, \cdots, N,$$
$$v_i(t) \neq 0, \ t > 0, \ i = 0, 1, 2, \cdots, N$$

and have

$$f_1(t) = v_N(t)$$
$$f_i(t) = v_N(t) \int_0^t v_{N-1}(t_1)dt_1 \int_0^t v_{N-2}(t_2)dt_2 \int_0^{t_2} \cdots$$
$$\cdots \int_0^t v_{N-i+1}(t_{i-1})dt_{i-1}, \ i = 1, 2, \cdots, N.$$

Also we have

$$g_i(u) = (-1)^{N-i} v_0(u) \int_0^u v_1(u_1)du_1 \int_0^{u_1} v_2(u_2)du_2 \int_0^{u_2} \cdots$$
$$\cdots \int_0^{u_{N-i-1}} v_{N-i}(u_{N-i})du_{N-1}, \ i = 1, 2, \cdots, N-1,$$
$$\cdots g_N(u) = v_0(u).$$

The operator L_t itself can be expressed in the form

$$L_t = \frac{1}{v_0(t)} D \frac{1}{v_1(t)} D \cdots \frac{1}{v_{N-1}(t)} D \frac{1}{v_N(t)} \qquad (7.1.10)$$

which has briefly been discussed in Section 4.2.

We continue to have observation from somewhat different viewpoint. The canonical kernel of a general N-ple Markov Gaussian process is a Goursat kernel. The f_i's and the g_i's have had different analytic properties. If, in particular, the process is N-ple Markov in the *restricted sense*, g_i's are not only linearly independent but have the same property as f_i's (see the expression in terms of the functions v_i's and exchange v_i with v_{N-i}). So, f_i's and g_i's have the *same* analytic properties.

Note 1. From now on, we will use the notation $(X(t), L_t, R)$ for a triple of a Gaussian Markov process (in a restricted sense) $X(t)$, the corresponding differential operator L_t and the Riemann function $R(t, u)$.

Note 2. Since L_t is a local operator, we may restrict t to be in a finite interval. However, in the stationary case we take the whole R^1 through which t runs.

7.2 Duality for multiple Markov Gaussian processes in the restricted sense

As is expressed in the previous section, we can see a duality between the two systems $\{f_i(t)\}$ and $\{g_i(t)\}$. Note that, in the case of $\{f_i\}$ and $\{g_i\}$ having the same analytic properties, where the expressions are obtained in the dual manner.

Theorem 7.4 *If a triple $(X(t), L_t, R), t \in [0,1]$, is given, then the N-ple Markov process $X^*(t)$ can be constructed such that $(X^*(t), L_t^*, R^*), t \in [0,1]$, is a dual system to $(X(t), L_t, R)$ in the sense that*

i) L_t^* is the formal adjoint of L_t,

ii) R^* is the Riemann function of L_t^*,

iii) $X^*(t) = \int_t^1 R^*(t,u)\dot{B}(u)du$ is the solution of

$$L_t^* X^*(t) = \dot{B}(t),$$

where $\dot{B}(t)$ is the same $\dot{B}$ given in (7.1.9) up to sign.
In addition,

$$(X^*)^*(t) = X(t).$$

Proof. i) The formal adjoint L_t^* of L_t, which is given by (7.1.10), is of the form (ignoring $(-1)^N$)

$$L_t^* = \frac{1}{v_N(t)} D \frac{1}{v_{N-1}(t)} D \cdots \frac{1}{v_1(t)} D \frac{1}{v_0(t)}. \tag{7.2.1}$$

Namely, for $f, g \in S$, (S : Schwartz space)

$$(L_t f, g) = (f, L_t^* g), \ (\cdot, \cdot) : \text{inner product} \tag{7.2.2}$$

We see that

$$L_t f_i(t) = 0, \ L_u^* g_j(u) = 0 \tag{7.2.3}$$

ii) There exists a Riemann function of L_t^*, let it be denoted by $R^*(t, u)$, i.e. $L_t^* R^*(t, u) = 0$.

iii) Construct $X^*(t)$ as

$$X^*(t) = \int_t^1 R^*(t, u) \dot{B}(u) du.$$

Then we have

$$L_t^* X^*(t) = \dot{B}(t).$$

Here $\dot{B}(t)$ is the same $\dot{B}$ as is given in (7.1.9) up to sign.

From the construction of $X^*(t)$, it is easy to see that

$$(X^*)^*(t) = X(t),$$

since $(L_t^*)^* = L_t$.

■

We can easily prove the following proposition with the same idea in the stationary case as above. The time interval is taken to be $(-\infty, \infty)$.

Proposition 7.1 *If $X(t), t \in (-\infty, \infty)$ is a stationary N-ple Markov in a restricted sense such that*

$$X(t) = \int_{-\infty}^t R(t-u) \dot{B}(u) du, \tag{7.2.4}$$

then there exists an ordinary differential operator L_t of order N with constant coefficients such that

$$L_t X(t) = \dot{B}(t)$$

and the dual process $X^*(t)$ is defined as

$$X^*(t) = \int_t^\infty R^*(t-u)\dot{B}(u)du, \tag{7.2.5}$$

where $R^*(u) = R(-u)$.

7.3 Uniformly multiple Markov processes

We will introduce a uniformly multiple Markov property[118] which satisfies somewhat stronger assumption than the multiple Markov processes discussed in the previous section. We require uniformity of dependence in time. With this property of g_i's in (7.1.8) are much restricted. The time parameter space is taken to be $[0,\infty)$.

Assume that $X(t), t \in [0,\infty)$ has the canonical representation of the form

$$X(t) = \int_0^t F(t,u)\dot{B}(u)du.$$

Take an interval $[a,b] \subset [0,\infty)$, and define the restriction $X(t;a,b)$ of $X(t)$ by

$$X(t;a,b) = X(t) - E\left(X(t)|\mathcal{B}_a(X)\right),\ a \le t \le b, \tag{7.3.1}$$

where $\mathcal{B}_a(X)$ is the σ-field generated by $\{X(\tau), \tau \le a\}$.

Definition 7.4 If a Gaussian process $X(t;a,b), t \in [a,b]$ is N-ple Markov for any $a, b \in R$ such that $0 < a < b < \infty$, then $X(t)$ is called a uniform N-ple Markov Gaussian process.

Obviously a stationary N-ple Markov Gaussian process is uniformly N-ple Markov.

Assume that $X(t)$ has canonical representation

$$X(t) = \int_0^t F(t,u)\dot{B}(u)du,\ t \ge 0, \tag{7.3.2}$$

where $F(t,u)$ is a Goursat kernel. That is

$$F(t,u) = \sum_{i=1}^N f_i(t)g_i(u). \tag{7.3.3}$$

We see that

$$X(t;a,b) = \int_a^t \left(\sum_{i=1}^N f_i(t) g_i(u) \right) \dot{B}(u) du, \; t \geq 0, \tag{7.3.4}$$

is the canonical representation. It is important to note that for any choice of [a,b], we can choose f_i, g_i to be those in (7.3.3).

Proposition 7.2 *The uniform Markov property asserts that for any interval $[a,b] \subset [0,\infty)$, g_i's restricted to $[a,b]$ are linearly independent in any $L^2([a,b])$.*

Remark 7.1 *The functions f_i and g_i have similar analytic properties, but to be able to claim the same property we have to consider the generalized Gaussian process. This will be discussed in the following section.*

7.4 The N-ple Markov property of a generalized Gaussian process

In this section we will discuss the N-ple Markov property for a generalized Gaussian process with the following subjects.

1. To determine a class of multiple Markov Gaussian processes, indeed generalized Gaussian processes, and form its representation.
2. To show that to each of them we can form a *dual* generalized Gaussian multiple Markov process with the same order of Markov property.
3. By taking generalized process, we can partly break the fetters of checking differentiability of functions involved.

Assume that the space E is σ-Hilbert and nuclear. More precisely E is topologized with respect to countably many Hilbertian norms $\| \, \|_n$ such that E_n, the completion of E by $\| \, \|_n$ is a Hilbert space and that the injection $E_n \mapsto E_{n-1}$ is of Hilbert-Schmidt type.

$$E = \bigcap_n E_n.$$

We take the basic probability space to be $(E^*, \mathcal{B}, \mu)$, E^* being the dual space of E.

To fix the idea, assume that E_n is the Sobolev space $K^n(R^1)$ of order n over R^1. We are now ready to define a generalized Gaussian process $X(\xi, x), \xi \in E$.

Assume that

(i) Each $X(\xi, x)$ is a random variable on $(E^*, \mathcal{B}, \mu)$, and $X(\xi, x), \xi \in E$ forms a Gaussian system with $E(X(\xi)) = 0$ for every $\xi \in E$.

(ii) $X(\xi)$ $(= X(\xi, x))$ is linear and strongly continuous (in $(L^2) = L^2(E^*, \mu)$ topology) in $\xi \in E$.

Let $\mathcal{B}_t(X)$ be the σ-field $(\subset \mathcal{B})$ generated by $\{X(\xi), supp(\xi) \subset (-\infty, t]\}$ and let $\mathcal{B}(X) = \bigvee_t \mathcal{B}_t(X)$. Define Hilbert spaces

$$L_t(X) = L^2(\mathcal{B}_t(X), \mu) \text{ and } L(X) = L^2(\mathcal{B}(X), \mu).$$

Again assume that

(iii) $L(X)$ is separable,

(iv) $\bigcap L_t(X) = \{0\}$ (purely non-deterministic).

With these assumptions we claim that the projection $E(t)$ such that

$$E(t) : L(X) \mapsto L_t(X)$$

is defined, and $\{E(t), t \in R\}$ forms a resolution of identity. We can therefore appeal to the Hellinger-Hahn theorem to have a direct sum:

$$L(X) = \bigoplus_n \mathcal{S}_n,$$

where $\mathcal{S}_n$ is a cyclic subspace obtained in the form

$$\mathcal{S}_n = \text{Span}\{dE(t)Y_n, t \in R\}$$

for some Y_n's in $L(X)$. In addition, there is a relaionship

$$d\rho_{n+1}(t) << d\rho_n(t),$$

where $d\rho_n(t) = \|dE(t)Y_n\|^2$.

Then, $N = \max\{n; d\rho_n \text{ is non-zero measure}\}$ is the multiplicity of $X(\xi)$.

There is one more assumption :

(v) $N = 1$ and $d\rho_1$ is equivalent to the Lebesgue measure.

With this assumption the additive Gaussian process $E(t)Y_1, t \in R$, may be taken to be a Brownian motion, denoted by $B(t)$.

We are now ready to define multiple Markov property. For this purpose, we require somewhat stronger condition (A) and (B) instead of (i) and (ii).

(A) $X(\xi)$ is a member of $\mathcal{H}_1^{(-N)}$, continuous in ξ, where $\mathcal{H}_1^{(-N)}$ is the extension of $\mathcal{H}_1$ such that $\mathcal{H}_1^{(-N)}$ is isomorphic to $K^{(-N)}(R^1)$, dual space of Sobolev space $K^N(R^1)$. (For details see [57], section 2.6.)

(B) $X(\xi, x)$ is continuous in $\xi \in K^N(R^1)$ for almost all x.

Definition 7.5 Under the general assumptions, $X(\xi)$ is an N-ple Markov generalized Gaussian process if for any fixed t_0 and for any linearly independent ξ_i's with $supp(\xi_i) \subset [t_0, \infty)$,

i) $\{E(X(\xi_i)|\mathcal{B}_{t_0}(X)), i = 1, 2, \cdots, N\}$ are linearly independent and

ii) $\{E(X(\xi_i)|\mathcal{B}_{t_0}(X)), i = 1, 2, \cdots, N+1\}$ are linearly dependent,

where $\mathcal{B}_{t_0}(X)$ is the σ-field generated by $X(\xi)$ with $supp(\xi) \subset (-\infty, t_0]$.

We start with the particular case $N = 1$.

Theorem 7.5 *Suppose $X(\xi)$ is 1-ple Markov. Then there exists $f \in K^{(-1)}(R^1)$ and $g^t \in K^{(-1)}(R^1)$ such that*

$$E(X(\xi)|\mathcal{B}_t(X)) = \langle f, \xi \rangle U_t, \tag{7.4.1}$$

where $supp(\xi) \subset (t, \infty)$ and $U_t = \int g^t(u)\dot{B}(u)du \in \mathcal{H}_1^{(-1)}$.

Proof. Let t be fixed. For $\xi_1, \xi_2 \in K^1$, $E(X(\xi_1)|\mathcal{B}_t(X))$ and $E(X(\xi_2)|\mathcal{B}_t(X))$ are linearly dependent, so that there exists a non-zero function $\varphi(\xi_1, \xi_2)$ such that

$$E(X(\xi_1)|\mathcal{B}_t(X)) = \varphi(\xi_1, \xi_2)E(X(\xi_2)|\mathcal{B}_t(X)). \tag{7.4.2}$$

Obviously $\varphi(\xi, \xi) = 1$ and $\varphi(\xi_1, \xi_2) = \varphi(\xi_2, \xi_1)^{-1}$.

For any ξ_3 supported by $[t, \infty)$, we have

$$E(X(\xi_2)|\mathcal{B}_t(X)) = \varphi(\xi_2, \xi_3)E(X(\xi_3)|\mathcal{B}_t(X)) \tag{7.4.3}$$

and

$$E(X(\xi_1)|\mathcal{B}_t(X)) = \varphi(\xi_1, \xi_3)E(X(\xi_3)|\mathcal{B}_t(X)). \tag{7.4.4}$$

Combining (7.4.2), (7.4.3) and (7.4.4), we have

$$\varphi(\xi_1, \xi_3) = \varphi(\xi_1, \xi_2)\varphi(\xi_2, \xi_3). \tag{7.4.5}$$

Choose any fixed ξ, say $\xi = \xi_2$, to have

$$\varphi(\xi_1, \xi_3) = \frac{\varphi(\xi_1, \xi)}{\varphi(\xi_3, \xi)}. \tag{7.4.6}$$

Then we may simply write

$$\varphi(\xi_1, \xi_3) = \frac{\varphi(\xi_1)}{\varphi(\xi_3)}. \tag{7.4.7}$$

Hence equation (7.4.4), taking $\xi_1 = \xi$, becomes

$$E(X(\xi)|\mathcal{B}_t(X)) = \frac{\varphi(\xi)}{\varphi(\xi_3)}E(X(\xi_3)|\mathcal{B}_t(X)) \tag{7.4.8}$$

which can be written as

$$E\left(X(\xi)|\mathcal{B}_t(X)\right) = \varphi(\xi)U_t(\xi_3),$$

where

$$U(t, \xi) = \frac{E(X(\xi)|\mathcal{B}_t(X))}{\varphi(\xi)},$$

with a note that φ never vanishes.

It can be easily seen from the equality (7.4.2) that $U(t, \xi)$ does not depend on ξ. Thus we will write $U(t)$ instead of $U(t, \xi)$. Also, $U(t)$ is a conditional expectation under the condition that it is $\mathcal{B}_t(X)$-measurable, so that it is expressed in the form $\langle g^t, \dot{B}\rangle$ with $g^t \in K^{(-1)}((-\infty, t])$ as a member of $\mathcal{H}_1^{(-1)}$ which is $\mathcal{B}_t(X)$-measurable.

For *supp*$(\xi) \subset [t, \infty)$, there exists, according to assumption A.1, a generalized function $g \in K^{(-1)}(R^1)$ such that

$$U(\tau) = \langle g(\tau, \cdot), \dot{B}(\cdot)\rangle$$

which is additive in $\mathcal{H}_1^{(-1)}$.

We further prove that, by using the fact

$$E(\cdot|\mathcal{B}_{t'}(X)) = E(E(\cdot|\mathcal{B}_t(X))|\mathcal{B}_{t'}(X)), \quad (t > t'),$$

the restriction of $g^t(u)$ to $(-\infty, t']$ coincides with $g^{t'}(u)$. Hence we conclude that there exists a generalized function $g(u)$ which is locally compact in $K^{(-1)}(R^1)$ such that $g^t(u)$ is a restriction of g to $(-\infty, t]$.

Then, we have

$$U(t) = \langle g^t, \dot{B} \rangle.$$

Since $\varphi(\xi)$ can be written as (f, ξ) with $f \in \widehat{K^{(-1)}(R^1)}$, we have

$$E\left(X(\xi)|\mathcal{B}_t(X)\right) = (f, \xi)U(t).$$

Hence the theorem is proved.

∎

7.5 Representations of multiple Markov generalized Gaussian processes

The final goal of this chapter, to be done in this section, is that Markov property, if extended to the class of generalized Gaussian process with the canonical representation, always guarantees the existence of its dual process.

We now come to the general case, i.e. N-ple Markov. Assume that multiplicity of X is 1. Then we have

$$\mathcal{B}_t(X) = \mathcal{B}_t(\dot{B}),$$

(i.e. the existence of canonical representation for ordinary Gaussian process). Under the assumptions (A) and (B) we can prove the following theorem.

Theorem 7.6 *Let $\{\xi_i\}$ be linearly independent. Suppose $X(\xi)$ is N-ple Markov. Then there exist two systems $\{f_i; 1 \le i \le N\}$ and $\{g_i; 1 \le i \le N\}$ such that each of them involves N linearly independent functions in $\widehat{K^{(-N)}(R^1)}$ and that $det(\langle f_i, \xi_j \rangle) \neq 0$. Further we have*

$$E\left(X(\xi)|\mathcal{B}_t(X)\right) = \sum_1^N \langle f_i, \xi \rangle U_i^t, \qquad \xi \in E \tag{7.5.1}$$

where $U_i^t = \langle g_i^t, \dot{B} \rangle$ and g_i^t being the restriction of g_i to $(-\infty, t]$.

Proof. The idea is the same as in Theorem 7.5. Let t be fixed. Assume that $supp(\xi_j) \subset [t, \infty), 1 \leq j \leq N$, and ξ_i's are linearly independent. Then $E(X(\xi_j)|\mathcal{B}_t(X))$, $j = 1, 2, \cdots N$ are linearly independent. Write $\xi_{N+1} = \zeta$. Then, we have, as a generalization of (7.4.2),

$$E(X(\zeta)|\mathcal{B}_s(X)) = \sum_{j=1}^{N} \varphi_j(\zeta; \xi_1, \cdots, \xi_N) E(X(\xi_j)|\mathcal{B}_s(X)), \tag{7.5.2}$$

where $supp(\zeta) \subset [s, \infty), s < t$.

Using the notation $\boldsymbol{\xi}$ for the vector $(\xi_1, \cdots, \xi_N)$, it is written as

$$E(X(\zeta)|\mathcal{B}_t(X)) = \sum_{j=1}^{N} \varphi_j(\zeta; \boldsymbol{\xi}) E(X(\xi_j)|\mathcal{B}_t(X)).$$

Then, we have for $\tau < s < t$ and $supp(\eta_j) \subset [\tau, \infty), 1 \leq j \leq N$,

$$\varphi(\zeta; \boldsymbol{\eta}) = \sum_{j=1}^{N} \varphi_j(\zeta; \boldsymbol{\xi}) \varphi(\xi_j; \boldsymbol{\eta}).$$

Take N different ζ_j's and denote $\{\varphi(\zeta_j; \xi_1, \cdots, \xi_N), j = 1, \cdots, N\}$ by $\boldsymbol{\varphi}(\boldsymbol{\zeta}\, ; \boldsymbol{\xi})$.

Then, we have

$$\boldsymbol{\varphi}(\boldsymbol{\zeta}; \boldsymbol{\eta}) = A(\boldsymbol{\zeta}, \boldsymbol{\xi}) \boldsymbol{\varphi}(\boldsymbol{\xi}; \boldsymbol{\eta}),$$

where $A(\boldsymbol{\zeta}, \boldsymbol{\xi}) = [\varphi_j(\zeta_i; \boldsymbol{\xi})]_{i,j=1,\cdots,N}$.

Using the above equation, it is obtained that

$$\boldsymbol{\varphi}(\zeta; \boldsymbol{\eta}) = A(\boldsymbol{\zeta}, \boldsymbol{\xi}) \boldsymbol{\varphi}(\xi; \boldsymbol{\eta}) = A(\boldsymbol{\zeta}, \boldsymbol{\xi}) A(\boldsymbol{\xi}, \boldsymbol{\eta}') \boldsymbol{\varphi}(\boldsymbol{\eta}'; \boldsymbol{\eta})).$$

But

$$\boldsymbol{\varphi}(\zeta; \boldsymbol{\eta}) = A(\boldsymbol{\zeta}, \boldsymbol{\eta}') \boldsymbol{\varphi}(\eta'; \boldsymbol{\eta}),$$

thus, we have

$$A(\boldsymbol{\zeta}, \boldsymbol{\eta}') = A(\boldsymbol{\zeta}, \boldsymbol{\xi}) A(\boldsymbol{\xi}, \boldsymbol{\eta}') \tag{7.5.3}$$

where $supp(\eta'_j) \subset [\tau', \infty)$, for each j, and $\tau' < \tau$.

Consequently, we have

$$A(\boldsymbol{\zeta}, \boldsymbol{\xi}) = A(\boldsymbol{\zeta}, \boldsymbol{\eta}') A^{-1}(\boldsymbol{\xi}, \boldsymbol{\eta}') \tag{7.5.4}$$

which does not depend on η'. So we can write as

$$A(\boldsymbol{\zeta},\boldsymbol{\xi})) = A(\boldsymbol{\zeta})A^{-1}(\boldsymbol{\xi}). \tag{7.5.5}$$

On the other hand, taking ζ_j, for ζ in the relation (7.5.2), we have

$$E(X(\zeta_j)|\mathcal{B}_t(X)) = \sum_{j=1}^{N} \varphi_k(\zeta_j;\boldsymbol{\xi})E(X(\xi_j)|\mathcal{B}_t(X)),$$

for $j = 1, \cdots, N$.

Denoting $(E(X(\zeta_1)|\mathcal{B}_t(X)), \cdots, E(X(\zeta_N)|\mathcal{B}_t(X)))$ by $\boldsymbol{E}(\boldsymbol{X}(\boldsymbol{\zeta})|\boldsymbol{\mathcal{B}}_t(\boldsymbol{X}))$, we have

$$\boldsymbol{E}(\boldsymbol{X}(\boldsymbol{\zeta})|\boldsymbol{\mathcal{B}}_t(\boldsymbol{X})) = A(\boldsymbol{\zeta},\boldsymbol{\xi})\boldsymbol{E}(\boldsymbol{X}(\boldsymbol{\xi})|\boldsymbol{\mathcal{B}}_t(\boldsymbol{X})). \tag{7.5.6}$$

Let

$$\boldsymbol{U}^t(\boldsymbol{\zeta}) = (U_1^t(\boldsymbol{\zeta}), \cdots, U_N^t(\boldsymbol{\zeta})) \tag{7.5.7}$$

such that

$$\boldsymbol{U}^t(\boldsymbol{\zeta}) = A^{-1}(\boldsymbol{\zeta})\boldsymbol{E}(\boldsymbol{X}(\boldsymbol{\zeta})|\boldsymbol{\mathcal{B}}_t(\boldsymbol{X})),$$

where $supp(\zeta_j) \cap [\tau, \infty) \neq \emptyset$, for each j and $\tau < t$.

Using the relation (7.5.5) and (7.5.6), we have

$$\begin{aligned}\boldsymbol{U}^t(\boldsymbol{\zeta}) &= A^{-1}(\boldsymbol{\zeta})\boldsymbol{E}(\boldsymbol{X}(\boldsymbol{\zeta})|\boldsymbol{\mathcal{B}}_t(\boldsymbol{X}))\\ &= A^{-1}(\boldsymbol{\zeta})A(\boldsymbol{\zeta},\boldsymbol{\xi})\boldsymbol{E}(\boldsymbol{X}(\boldsymbol{\xi})|\boldsymbol{\mathcal{B}}_t(\boldsymbol{X}))\\ &= A^{-1}(\boldsymbol{\zeta})A(\boldsymbol{\zeta})A^{-1}(\boldsymbol{\xi})\boldsymbol{E}(\boldsymbol{X}(\boldsymbol{\xi})|\boldsymbol{\mathcal{B}}_t(\boldsymbol{X}))\\ &= \boldsymbol{U}^t(\boldsymbol{\xi}).\end{aligned}$$

That is, $\boldsymbol{U}^t$ is independent of $\boldsymbol{\zeta}$ and so is U_j^t.

It can be seen that each U_j^t is in $\mathcal{H}_1^{(-N)}$ and "additive" and that U_j^t's are linearly independent.

So it can be expressed as

$$U_j^t = \langle g_j^t, \dot{B}\rangle$$

where $g_j^t \in \widehat{K^{-N}(R^1)}$.

Since $\varphi_j(\xi; \xi_1, \cdots, \xi_N)$ in (7.5.2) is a linear function of $\xi_1, \cdots, \xi_n$, it can be written as $\langle f_j, \xi \rangle$, with $f_j \in \widehat{K^{(-N)}(R^1)}$.

Thus the theorem is proved.

■

Definition 7.6 If N-ple Markov property holds for a generalized Gaussian process $X(\xi)$ in any time interval, then $X(\xi)$ is called uniformly N-ple Markov generalized process.

Theorem 7.7 *If a generalized Gaussian process $X(\xi)$ is uniformly N-ple Markov process then it is N-ple Markov and f_i are linearly independent in $K^{(-N)}(R^1), det(\langle g_i, \xi_j \rangle) \neq 0$, for ξ_j's being linearly independent.*

The above theorem implies that two systems $\{f_i\}$ and $\{g_j\}$ of $K^{-N}(R^1)$ functions have the same property as far as linearly independence is concerned. We now finally come to our main theorem.

Theorem 7.8 *If the generalized Gaussian process $X(\xi)$ is a uniformly N-ple Markov process then the dual process of $X(\xi)$ exists and is expressed in the form*

$$E(X^*(\xi)|\mathcal{B}_t(X)) = \sum_{i=1}^{N} \langle g_i, \xi \rangle U_i^*(t) \tag{7.5.8}$$

where $U_i^*(t) = \langle f_i^t, \dot{B} \rangle$.

Note that stationary N-ple Markov process as well as strictly Markov process is uniformly N-ple Markov process.

7.6 Stationary N-ple Markov generalized Gaussian process

We assume that $X(\xi)$ is stationary N-ple Markov generalized Gaussian process and is purely non-deterministic.

Theorem 7.9 *Under the above assumption, $\{f_i\}$ satisfies the following equation*

$$E(X(\xi)|\mathcal{B}_t(X)) = \sum_{1}^{N} \langle f_i, \xi \rangle U_i(t) \tag{7.6.1}$$

and is a fundamental system of solutions of a differential equation

$$L_t f_i = 0,\ L_t = \sum_1^N a_i(\frac{d}{dt})^{N-i}, \tag{7.6.2}$$

where $U_i(t) = \langle g_i, \dot{B}\rangle$ *and* a_j*'s are constants. In addition, the dual process of* $X(\xi)$ *satisfies*

$$E(X^*(\xi)|\mathcal{B}_t(X)) = \sum_{j=1}^N \langle g_j, \xi\rangle U_j^*(t) \tag{7.6.3}$$

where $U_j^*(t) = \langle f_j^*, \dot{B}\rangle$ *with* $\{f_j^*\}$ *being a fundamental system of solutions of a differential equation*

$$L_t^* f_j^* = 0,\ L_t^* = \sum_1^N (-1)^{N-j} a_j(\frac{d}{dt})^{N-j}. \tag{7.6.4}$$

Proof uses the same technique as in the literature[21] to have the operator L_t.

Theorem 7.10 *The stationary* N*-ple generalized Gaussian process is in the space* $\boldsymbol{S}_N$ *in Ito's sense.*

Proof. The spectral density $f(\lambda)$ of the stationary N-ple generalized Gaussian process is a rational function which is of the form

$$f(\lambda) = \frac{Q(\lambda^2)}{P(\lambda^2)},$$

where P and Q are polynomials in λ^2 and degrees of P and Q are N and at most $2N-1$, respectively. Thus, using Itô's classification (see the Appendix) we obtain the assertion.

■

Example 7.2 Let $X(t)$ be a stationary double Markov generalized Gaussian process with the spectral density function :

$$f(\lambda) = \frac{\lambda^2(c^2+\lambda^2)}{(a^2+\lambda^2)(b^2+\lambda^2)}, \tag{7.6.5}$$

where a, b, c are positive different real numbers.

Set

$$Y(t) = \int_{-\infty}^{t} \left(\frac{c-a}{b-a} e^{-a(t-u)} + \frac{b-c}{b-a} e^{-b(t-u)} e^{-b(t-u)} \right) \dot{B}(u) du.$$

This expression is the canonical representation. Since

$$F(0) = \frac{c-a}{b-a} + \frac{b-c}{b-a} \neq 0,$$

$Y(t)$ is not differentiable. So $\frac{d}{dt}Y(T) = Y'(t)$ is a generalized Gaussian process and the spectral density is equal to $f(\lambda)$ given by equation (7.6.5). Namely, $Y'(t)$ is the same process as $X(t)$ which is given in the beginning of this example.

The dual process of $Y(t)$ is given by

$$Y^*(t) = \int_{t}^{\infty} \left(\frac{c-a}{b-a} e^{a(t-u)} + \frac{b-c}{b-a} e^{b(t-u)} \right) \dot{B}(u) du.$$

The process $X(t)$ with the spectral density function (7.6.5) is a velocity of the dynamics.

If the spectral distribution function is

$$f(\lambda) = \frac{c^2 + \lambda^2}{(a^2 + \lambda^2)(b^2 + \lambda^2)}, \tag{7.6.6}$$

the corresponding process is an ordinary stochastic process which is formed by Lee-Wiener network (see the literature[27]).

We can see that

$$X(t) = Y'(t)$$

and it is a generalized process belongs to Ito's class $\mathbf{S_2}$.

For the representation of the kernel we have to specify the signs of a and b.

Example 7.3 Let $X(t)$ be a N-ple Markov process in Doob's sense.

i) Let $u(t)$ be not in the vector space spanned by the functions belonging to the fundamental system of solutions of the differential equation

$$L_t f = 0,$$

where L_t is the nth order differential operator such that

$$L_t X(t) = \dot{B}(t), \ X(0) = 0.$$

Then

$$\frac{d}{dt}\frac{1}{u(t)}X(t)$$

is defined and it is still N-ple Markov, since u cannot be a function of Frobenius formula. We see that differential operator does not always effect to decrease the order of Markov property.

ii) Another example obtained by a multiplication by a function $\varphi(t)$ to define

$$Y(t) = \varphi(t)X(t),$$

where $\varphi(t)$ is non-zero and non-differentiable on any interval. The product makes sense and we can prove that $Y(t)$ is N-ple Markov generalized Gaussian process although the regularity of sample functions are quite different from those of $X(t)$.

7.7 Remarks on the definition of multiple Markov property

A significance of taking the class of generalized Gaussian process to define multiple Markov process can be seen by comparing the following interpretations [I] and [II].

[I] An N-ple Markov $X(t)$ in the restricted sense

Definition uses an operator L_t which is an N-th order differential operator. We must assume $X(t)$ to be $(N-1)$ times differentiable, and the N-th derivative is in the Hida distribution. Anyway, in white noise analysis we do assume N-times differentiable. This number N is in agreement with the order of Markov property. The Riemann function defines the dual process. Elementary differential calculus allows us to obtain duality. (Compare with Section 7.6, III.)

[II] An N-ple Markov generalized Gaussian process $X(\xi)$

What we have discussed in this paper emphasizes the dependence and analytic property do not head the cast. In order to define ***N*-*ple Markov property***, (implicitly assume a possibility to get a dual process) it is quite reasonable to assume not differentiability but continuity in ξ which is in *the Sobolev space of degree* $\boldsymbol{N}$.

In our case the random function $X(\xi)$ is assumed to be in Hida distribution space of order $(-N)$. This plays some role of Riemann function which does not exist in our case of generalized Gaussian processes.

The class which we have determined seems to be **optimal**, since we do not assume the differentiability heavily.

[III] A bridge to the group $GL(N, R)$

We have established the uniqueness of the canonical representation. The kernel function $F(t, u)$ is unique up to sign. For an N-ple Markov Gaussian process, its canonical kernel is a Goursat kernel is expressed as (7.1.8). Note that each f_i is not unique, the same for g_i. Use the terminology in Section 7.1. III, and write $\boldsymbol{F}(t) = (F_1(t), \cdots, F_N(t)), \boldsymbol{U}(t) = (U_1(t), \cdots, U_N(t))$. If we take a matrix A in $SL(N, R)$, then $A\boldsymbol{U}(t))$, and

$$X(t) = (A^* \boldsymbol{F}(t), A\boldsymbol{U}(t))$$

holds. To guarantee the above equality we have a freedom to take a nonsingular matrix B to have $(B^{-1})^*(A^* \boldsymbol{F}(t), BA\boldsymbol{U}(t))$. Indeed, this is the only possible alternative expression of the canonical representation. Thus, we claim, writing $\boldsymbol{G} = (g_1, \cdots, g_N)$:

Proposition 7.3 *Two expressions* $(\boldsymbol{F}_i, \boldsymbol{G}_i), i = 1, 2,$ *of the canonical kernels of an* N*-ple Markov Gaussian process are linked by*

$$\boldsymbol{F}_2 = (B^{-1})^* A^* \boldsymbol{F}_1, \boldsymbol{G}_2 = BA\boldsymbol{U}(t),$$

where $A \in SL(N, R)$ *and* B *is nonsingular.*

The method we have discussed is independent of t.

Chapter 8

Classification of Noises

A noise is a system of idealized elemental random variables which is a system of idealized (or generalized), independent, identically distributed random variables parametrized by an ordered set.

If the parameter set is discrete, say $\boldsymbol{Z}$, then we simply have i.i.d. random variables $\{X_n, n \in \boldsymbol{Z}\}$, which is easily dealt with. While the parameter set is continuous, say R^1, then we have to be careful with the definition and analysis of functionals of the noise.

In this chapter, we shall find possible noises under reasonable assumptions and give a classification of them. We consider only noises depending on the continuous parameter. They are classified by the type. It is noted that two noises are said to be the same type if and only if the probability distributions are of the same type.

We classify the well-known noises according to the choice of parameter. A noise usually consists of continuously many *idealized* random variables, and they are parametrized by an ordered set.

Remark 8.1 *We say idealized random variables, because if they were ordinary random variables as many as continuum, then the probability distribution of the system in question could not be an abstract Lebesgue space; that is, it is impossible to carry on the calculus with the help of the Lebesgue type integral. It is therefore necessary and natural to have idealized (generalized) random variables.*

Noises can be classified as follows.

a) Time dependent noises

(Gaussian) White Noise ; $\dot{B}(t)$

Poisson Noise : $\dot{P}(t)$,

They are well known, so that there is no need to explain.

b) Space dependent noise $P'(u), u > 0$.

This is the system we are going to discuss in this chapter.

8.1 The birth of noises

We wish to find possible noises by approximation under reasonable assumptions. We take unit interval I as a representative of continuous parameter set. Approximation should be uniform in $u \in I$, and the idealized variables should be i.i.d. so is for the approximating variables. There are two cases where one is that approximating variables are having a finite variance, and another one is that they are atomic.

Consider a linear parameter. To fix the idea we take the unit interval $I = [0, 1]$. In case the **reduction** is successful, we are given a noise with parameter set I.

Case I. Let $\boldsymbol{\Delta}^n = \{\Delta_j^n, 1 \le j \le 2^n\}$ be the partition of I. Assume that $|\Delta_j^n| = 2^{-n}$.

To each subinterval Δ_j^n we associate a random variable X_j^n. Assume that $\{X_j^n\}$ are i.i.d. (independent identically distributed) with mean 0 and finite variance v.

Let n be large. Then, we can appeal to the central limit theorem to have a standard Gaussian distribution $N(0, 1)$ as the limit of the distribution of

$$S_n = \frac{\sum_1^{2^n} X_j^n)}{\sqrt{2^n v}}.$$

If we take subinterval $[a, b]$ of I, then the same trick gives us a Gaussian distribution $N(0, b-a)$, which is, for the later purpose, denoted by $N(a, b)$. The family

$$\mathcal{N} = \{N(a, b); 0 \le a, b \le 1\}$$

forms a consistent system of distributions, so that the family $\mathcal{N}$ defines a stochastic process. It is nothing but a Brownian motion $B(t), t \in [0, 1]$.

Case II. The interval I and its partition Δ^n are the same as in the case I. The independent random variables X_j^n are i.i.d., but all are subject to a simple probability distribution such that

$$P(X_j^n = 1) = p_n, \; P(X_j^n = 0) = 1 - p_n.$$

Let p_n be smaller as n is getting larger keeping the relation

$$2^n p_n = \lambda$$

for some positive constant $\lambda > 0$. Then, by the *law of small probabilities of Poisson* (this term comes from Lévy[80]) we are given the Poisson distribution $P(\lambda)$ with intensity λ.

Take a sub-interval $[a, b]$ of I. As in the case I, we can define a Poisson distribution with intensity $(b - a)\lambda$, denoted by $P([a, b], \lambda)$. We, therefore, have a consistent family of Poisson distributions $\mathcal{P} = \{P([a, b], \lambda)\}$. Hence $\mathcal{P}$ determines a Poisson process with intensity λ.

Here are the important notes.

1) We have a freedom to choose the constant λ whatever we wish, so far as it is positive.
2) The positive constant λ is the expectation of a random variable $P(\lambda) = P([0, 1], \lambda)$. It can be viewed as *scale* or *space* variable.

We now understand that a noise, which is taken to be a realization of the randomness due to reduction, can eventually create a space variable.

Case III. Next step of our study is concerned with a general new noise depending on a space parameter.

Keep random variables X_i^n as above and divide the sum S_n into partial sums :

$$S_n = \sum_j S_n^j,$$

where

$$S_n^j = \sum_{i=k_j+1}^{k_{j+1}} X_i^n,$$

with

$$1 = k_0 < k_1 < k_2 < \cdots < k_m = 2^n.$$

We assume that $k_{j+1} - k_j \to \infty$ as $n \to \infty$ for every j and that each ratio $\frac{k_{j+1}-k_j}{2^n}$ converges to $\frac{\lambda_j}{\lambda}$, respectively.

Theorem 8.1 *Let S_n^j, S_n, λ_j and λ be as above. Let $P(\lambda_j), 1 \le j \le m$ be mutually independent Poisson random variables with intensity λ_j, respectively. Then, S_n^j and S_n converges to $P(\lambda_j)$ and $P(\lambda)$ in law, respectively.*

Proof. It is easy by observing the characteristic functions $\varphi_n^j(z)$ and $\varphi_n(\boldsymbol{z})$ of S_n^j and S_n, respectively. The evaluation gives

$$\varphi_n^j(z_j) = (1 + \frac{\lambda_j}{(k_{j+1} - k_j)}(e^{iu_j z_j} - 1))^{n_j}, \tag{8.1.1}$$

and

$$\varphi_n(\boldsymbol{z}) = \prod_{j=1}^{m} \left(1 + \frac{\lambda_j}{(k_{j+1} - k_j)}(e^{iu_j z_j} - 1)\right)^{n_j}. \tag{8.1.2}$$

Thus, $\varphi_n^j(z)$ tends to $e^{\lambda_j(e^{iu_j z_j} - 1)}$ which is the characteristic function of $P(\lambda_j)$ and $\varphi_n(z)$ tends to

$$\varphi(\boldsymbol{z}) = \prod_{j=1}^{m} e^{\lambda_j(e^{iu_j z_j} - 1)}, \tag{8.1.3}$$

which is the characteristic function of $P(\lambda)$.

This proves the assertion.

■

Corollary 8.1 *The characteristic function $\varphi(z)$ of $P = \sum_k P(\lambda_k)$ is expressed in the form*

$$\varphi(\boldsymbol{z}) = e^{\sum_k \lambda_k(e^{iu_j z_j} - 1)}.$$

Before we come to the next topic, we pause to see the following facts.

The type of probability distributions

Definition 8.1 The probability distributions F and G are said to be the same type if there exist constants $a > 0$ and b such that $F(x) = G(ax + b)$.

Equivalently, we can say that the probability distributions of the random variables X and Y are the same type if there exist constants $\alpha > 0$ and β such that X and $\alpha Y + \beta$ have the same distributions.

All the Gaussian distributions are the same type (the exceptional Gaussian distribution $N(m, 0)$ is excluded).

Example 8.1 Let X have Gaussian $N(m, \sigma^2)$. Then it is the same type as $N(0, 1)$. Hence any two Gaussian distributions are the same type.

On the other hand, Poisson type distributions with different intensities are not the same type. This can be proved by the formula of characteristic function.

Example 8.2 Take the two Poisson distributions with intensity λ_1 and λ_2, respectively.

Compare two characteristic functions :

$$\varphi_1(z) = e^{\lambda_1(e^{iz}-1)},$$
$$\varphi_2(z) = e^{\lambda_2(e^{iz}-1)}.$$

These two functions of z cannot be exchanged by any affine transformation of z if $\lambda_1 \neq \lambda_2$. Thus we can see that Poisson distributions with different intensities are not the same type.

Note. We can say that, concerning the Poisson type,

1) There is a freedom to choose intensity arbitrary. Hence we can form, by the sum of i.i.d. random variables, continuously many Poisson type random variables of different type.

2) The intensity is a parameter, different from the time parameter, to be remarkable. The intensity is viewed as a space parameter. The above construction shows that it is additive in λ.

3) Multiplication by a constant to Poisson type variable keeps the type. The constant, however, can play the role of a *label*. So, take a constant $u = u(\lambda)$ as a *label* of the intensity λ. The function $u(\lambda)$ is therefore univalent. In view of this fact, we can form an inverse function $\lambda = \lambda(u)$ which is to be monotone.

With the remark made above, we change our eyes towards multi-dimensional view. Consider a vector consisting independent components

$$P(\lambda) = (P(\lambda_1), P(\lambda_2), \cdots, P(\lambda_n))$$

and its characteristic function

$$\varphi(\boldsymbol{z}) = \prod_{k=1}^{n} e^{\lambda_k (e^{iu_k z_k} - 1)},$$

where $\boldsymbol{z} = (z_1, \cdots, z_n)$.

Now it is requested to identify every component $P(\lambda_k)$, so that we give a label to each $P(\lambda_k)$, say different real number u_k to have $u_k P(\lambda_k)$.

Now let us have passage from integer $\{k\}$ to real $u > 0$, that is, passage from intensity λ_k and $u_k P(\lambda_k)$ to $\lambda(u)$ and $uP(\lambda(u))$, respectively.

Here we pause to focus our attention on the intensity $\lambda(u)$. As we have noted in Example 8.2, we formulate the following proposition.

Proposition 8.1 *Let $X(\lambda)$ be subject to a Poisson distribution with intensity λ. Then*

i) the probability distribution of $X(\lambda)$ and $uX(\lambda), u > 0$ are the same type, while
ii) $X(\lambda)$ and $X(\lambda')$ have distribution of different type.

Proof comes from the formulas of characteristic functions.

Theorem 8.2 *The characteristic function $\varphi(\mathbf{z})$ turns into a functional of ξ on a nuclear space E which is a dense subspace of $L^2([0, \infty))$, expressed in the form*

$$C^P(\xi) = \exp[\int \lambda(u)(e^{iu\xi(u)} - 1)du]. \tag{8.1.4}$$

What we have done is that starting from a higher dimensional characteristic function of Poisson type distribution, we have its limit $C^P(\xi)$.

8.2 A noise of new type

We are going to see a new noise. It is, of course, in line with the study of random functions using i.e.r.v.'s.

The idea is to choose another kind of parameter, say space variable, instead of time. Motivation comes from the decomposition of a compound Poisson process.

By the discussions in the previous section lead us to consider a functional $C^P(\xi)$, where the variable ξ runs through a certain nuclear space $E \subset C^\infty((0,\infty))$, say isomorphic to the Schwartz space $\mathcal{S}$;

$$C^P(\xi) = \exp[\int_{R^+} (e^{iu\xi(u)} - 1)\lambda(u)du], \tag{8.2.1}$$

where $R^+ = (0,\infty)$ and where $\lambda(u)du$ is a measure on $(0,\infty)$ to be specified so that $\lambda(u)$ positive a.e. and $\frac{u^2}{1+u^2}\lambda(u)$ is integrable.

Theorem 8.3 *Under these assumptions, the functional $C^P(\xi)$ is a characteristic functional. That is*

i) $C^P(\xi)$ is continuous in ξ, with a suitable choice of a nuclear space E.
ii) $C^P(0) = 1$.
iii) $C^P(\xi)$ is positive definite.

Proof. i) can be proved as follows.

We form a σ-Hilbert nuclear space E, on which $C^P(\xi)$ is continuous.

Take the Schwartz space S and introduce a relation $\sim$ on S in such a way that for ξ and $\eta \in S$, $\xi \sim \eta$ if and only is $\xi(u) = \eta(u)$ on $(0,\infty)$.

Then "$\sim$" satisfies the equivalence relations, so that we can form a factor space

$$S^+ = S/\sim .$$

Since S^+ is a factor space, the Schwartz topology is naturally reduced to S^+, and it is a σ-Hilbert nuclear space.

On $E = S^+$, we now prove that $C^P(\xi)$ is continuous.

Suppose $\xi_n \to \xi$ on E. It is easy to see that $C^P(\xi_n) \to C^P(\xi)$ if and only if $\int_0^\infty u\xi_n(u)\lambda(u)du \to \int_0^\infty u\xi(u)\lambda(u)du$.

This comes from the following evaluations.

$$
\begin{aligned}
&|\int_0^\infty (u\xi_n(u) - u\xi(u))\lambda(u)du| \\
&\leq \int_0^1 u|\xi_n(u) - \xi(u)|\lambda(u)du + \int_1^\infty u|\xi_n(u) - \xi(u)|\lambda(u)du \\
&\leq \int_0^1 \frac{|\xi_n(u) - \xi(u)|}{u} u^2\lambda(u)du + \int_1^\infty \sup_u u|\xi_n(u) - \xi(u)|\lambda(u)du \\
&\leq \int_0^1 \sup_{0\leq\theta\leq 1} |\xi_n'(\theta) - \xi'(\theta)|u^2\lambda(u)du + \int_1^\infty \sup_u u|\xi_n(u) - \xi(u)|\lambda(u)du
\end{aligned}
$$

By assumption on $\lambda(u)$,

$$
\int_0^1 u^2\lambda(u)du < \infty, \quad \int_1^\infty \lambda(u)du < \infty,
$$

we see that the last sum tends to 0 by using the Schwartz topology and the assertion holds.

ii) is obvious.

iii) comes from the fact that the characteristic function of Poisson type distribution $\exp[(e^{izu} - 1)\lambda] = \varphi(z;\lambda)$ is positive definite and $C^P(\xi)$ is expressed as a product of $\varphi(z;\lambda)$.

■

Applying the Bochner-Minlos theorem, we can see that there exists a probability measure ν^P on E^* such that

$$
C^P(\xi) = \int_{E^*} e^{i\langle x,\xi\rangle} d\nu^P(x). \tag{8.2.2}
$$

We introduce a notation $P'(u, \lambda(u))$ or write it simply $P'(u)$. We understand that ν^P-almost all $x \in E^*, x = x(u), u \in R^1$ is a sample function of a generalized stochastic process $P'(u), u \in R^+$.

Theorem 8.4 *The $\{P'(u), u \in R^+\}$ defines a system of i.e.r.v., that is a noise.*

Proof. Suppose ξ_1 and ξ_2 have disjoint support; let them be denoted by A_1 and A_2. Then, the integral in the expression (8.2.1) can be expressed as a sum

$$
\int_{A_1} (e^{iu\xi(u)} - 1)\lambda(u)du + \int_{A_2} (e^{iu\xi(u)} - 1)\lambda(u)du.
$$

This proves the equality

$$C^P(\xi_1 + \xi_2) = C^P(\xi_1)C^P(\xi_2).$$

This shows that $P'(u)$ has independent value at every point u. Also follow the atomic property and others.

■

The bilinear form $\langle P', \xi \rangle, \xi \in E$ is a random variable with mean

$$\int u\xi(u)\lambda(u)du.$$

It is, of course, not permitted to put $\xi = \delta(u_0)$ to say $E(P'(u_0)) = u_0\lambda(u_0)$.

The variance of $\langle P', \xi \rangle$ is

$$\int u^2\xi(u)^2\lambda(u)du.$$

Hence $\langle P', \xi \rangle$ extends to $\langle P', f \rangle$ under the condition $uf(u) \in L^2((0, \infty), \lambda(u)du)$.

If $\langle P', f \rangle - uf(u)$ and $\langle P', g \rangle - ug(u)$ are orthogonal in $L^2(\lambda(u)du)$, then $\langle P', f \rangle$ and $\langle P', g \rangle$ are uncorrelated. Thus, we can form a random measure and hence, we can define the space $\mathcal{H}_1(P)$ like $\mathcal{H}_1$ in the case of Gaussian white noise. The space $\mathcal{H}_1(P)$ can also be extended to a space $\mathcal{H}_1^{(-1)}(P)$ of generalized linear functionals of $P'(u)$'s. Note that there we can give an *identity* to $P'(u)$ for any u.

Our conclusion is that single noise $P'(u)$ with the parameter u can be found in $\mathcal{H}_1^{(-1)}(P)$.

Remark 8.2 *The case where the parameter u runs through the negative interval $(-\infty, 0)$ can be discussed in the similar manner. It is, however, noted that the single point mass at $u = 0$ is omitted.*

We now come to a next topic. Since u is supposed to be the space variable running through R_+, it is natural to consider the dilation of u

$$u \to au,$$

in place of the shift in the time parameter case, $t \in R$.

The $P'(u)$ is said to be self-similar if for any $a > 0$ there exists $d(a)$ such that

$$P'(au) \sim d(a)P'(u),$$

where $\sim$ means the same distribution.

Theorem 8.5 *If $P'(u)$ is self similar, then*

$$\lambda(u) = \frac{c}{u^{1+\alpha}},$$

where c is a positive constant and $0 < \alpha < 1$.

Proof is easy and is omitted.

The above case corresponds to the stable noise and α is the exponent of the noise.

Since we exclude $u = 0$, and since we may restrict our attention to the positive parameter set, i.e. $(0, \infty)$. We now consider the one-parameter group

$$\mathcal{G} = \{g_t, t \in R^1\}$$

of dilations, where

$$g_t : u \rightarrow g_t u = e^{at}u, \ a > 0.$$

There, we have

$$g_t g_s = g_{t+s}.$$

Remark 8.3

1) We can claim the same result for the case $1 \leq \alpha < 2$ by a slight modification of the characteristic functional.

2) The group $\mathcal{G}$ characterizes the class of the stable noises. More details shall be discussed in the forthcoming paper.

8.3 Invariance

Like white noise, we can see invariance with respect to dilation and reflection acting on $u \in (0, \infty)$.

1) Isotropic dilation

Since the parameter u runs through R_+, the dilation is the basic transformation.

$$x \in (-\infty, \infty) \longleftrightarrow e^x \in (0, \infty)$$
$$x \mapsto x + h \longleftrightarrow e^x \mapsto e^{x+h} = e^x e^h, \quad (\text{i.e. } u \mapsto au)$$
$$\text{i.e. Shift} \longleftrightarrow \text{Dilation}$$

$$g_a : u \mapsto au$$

is isotropic dilation.

Isotropic dilation group

Let

$$\mathcal{G} = \{g_a, g_a u = au\}.$$

There, we have

$$g_a g_b = g_{a+b},$$

so $\mathcal{G}$ is the group of dilations.

Remark 8.4 *The group $\mathcal{G}$ characterizes the class of the stable noises. The type of stable noise does not change under dilation group.*

2) Reflection with respect to the unit sphere S^1

Consider the reflection

$$u \longmapsto \frac{1}{u},$$

then

$$\lambda(u) = \frac{1}{u^{\alpha+1}} \longmapsto \lambda(\frac{1}{u}) = u^{\alpha+1}.$$

The reflection changes the intensity for a stable process as follows :

$$\lambda(u) \longmapsto \lambda(\frac{1}{u}) \cdot \frac{1}{u^3} = \frac{1}{u^{2-\alpha}} = \frac{1}{u^{(1-\alpha)+1}}.$$

That is, when α changes to $1-\alpha$, the range of α changes as

$$0 < \alpha < 1 \quad \longleftrightarrow \quad 1 > 1 - \alpha > 0.$$

Theorem 8.6 *The family $\{P'_\alpha(\lambda, u), 0 < \alpha < 1\}$ is invariant under the reflection*

$$\lambda(u) \longmapsto \lambda(\frac{1}{u}) \cdot \frac{1}{u^3},$$

where $\alpha \to 1 - \alpha$.

3) Inverse and reciprocal transformation of exponent

We first confer the motivation of discussing these transformations. That is the inverse transform of Brownian motion.

Let $B(t)$ be a Brownian motion, $M(t)$ be the maximum of Brownian motion, i.e. $M(t) = \max_{s \le t} B(t)$, and $T(y)$ be the inverse function of $M(t)$. Thus, we have

$$B(t) \longmapsto M(t) \longmapsto T(y).$$

Then, $T(y)$ is a stable process with exponent $\frac{1}{2}$.

Hence, we see that the exponent $\alpha = 2$ of the Brownian motion changes to the exponent $\alpha = \frac{1}{2}$.

Reciprocal transformation of exponent for new noise

Proposition 8.2 *By changing the exponent α of a new noise to its reciprocal $\frac{1}{\alpha}$, then the corresponding intensity is*

$$\lambda_{\frac{1}{\alpha}}(u) = (\lambda_\alpha(u))^{\frac{1}{\alpha}}.$$

Proof. Change $\alpha \longmapsto \frac{1}{\alpha} \Longrightarrow$ change the intensity $\lambda_\alpha \longmapsto \lambda_{\frac{1}{\alpha}}$

Then we have

$$\lambda_\alpha(u) = \frac{1}{u^{\alpha+1}} \longmapsto \lambda_{\frac{1}{\alpha}}(u) = \frac{1}{u^{\frac{1}{\alpha}+1}} = (\frac{1}{u^{\alpha+1}})^{\frac{1}{\alpha}}$$

Thus, we have

$$\lambda_{\frac{1}{\alpha}}(u) = (\lambda_\alpha(u))^{\frac{1}{\alpha}}.$$

■

For a general stable distribution

We now generalize this mapping $\alpha \longmapsto \frac{1}{\alpha}$ where α is the exponent of stable distribution.

Let $X_\alpha(t)$ be a stable process with exponent α and $T(y)$ be the inverse function of $M_\alpha(t) = \max_{s\le t} X_\alpha(s)$.

Theorem 8.7 *The probability distribution of $y^{-\alpha}T(y)$ is independent of y.*

Proof is given by Lévy[82].

Change the characteristic functional

$$C_\alpha^P(\xi) = \exp\left[\int_0^\infty (e^{iu\xi(u)} - 1 + iu\xi)\frac{1}{u^{\alpha+1}}du\right]$$

to

$$C_\beta^P(\xi) = \exp\left[\int_0^\infty (e^{iu\xi(u)} - 1)\frac{1}{u^{\beta+1}}du\right],$$

where $\alpha\beta = 1, 1 < \alpha < 2$.

Note that C_P^α involves the term of compensation $iu\xi(u)$.

For the Cauchy distribution, intensity does not change for every label.

8.4 Representation of linear generalized functionals of $P'(u)$'s

We shall focus our attention on the space of linear functionals of $P'(u)$.

Up to the common factor $C^P(\xi)$, we are given a representation of a linear function of Poisson noise expressed in the form

$$\int e^{iu\xi(u)} u\eta(u)\lambda(u)du.$$

By subtracting off the constant (expectation) we have

$$\int (e^{iu\xi(u)} - 1)u\eta(u)\lambda(u)du.$$

This is linear in η. We are, therefore, given a linear space

$$\mathcal{F}_1 = \text{span}\left\{ \int (e^{iu\xi(u)} - 1)u\eta(u)\lambda(u)du, \ \eta \in E \right\}, \tag{8.4.1}$$

which is isomorphic to

$$\mathcal{H}_1(P) = \text{span}\{\text{linear functionals of } P'(u)\}.$$

The $\langle x, \xi \rangle, \xi \in E$, is viewed as a sample of a random variable $\langle P', \xi \rangle$, the characteristic function of which is given by $\varphi^{P'}(z) = C^P(z\xi)$.

The bijection

$$\xi \Longleftrightarrow \langle P', \xi \rangle, \ \xi \in E$$

leads us to have

$$f \Longleftrightarrow \langle P', f \rangle, \ uf \in L^2((0, \infty), d\lambda),$$

where $d\lambda = \lambda(u)du$.

To establish a general theory of representation of nonlinear functionals of $P'(u)$, we can appeal to the theory of Reproducing Kernel Hilbert Space with kernel $C^P(\xi - \eta)$, or the $\mathcal{T}$-transform, which is an analogue of the Fourier transform. See the monograph[57].

By using the representation, we can rigorously define the random measure $p(du)$ (which was briefly mentioned before) such that for any Borel subset $B \subset (0, \infty)$ with finite measure $d\lambda$

$$p(B) = \langle P', \chi_B \rangle,$$

is defined, where χ_B is the indicator function of the Borel set B.

Stochastic integrals based on $p(du)$ are defined in the usual manner, and the collections of the stochastic integrals forms a Hilbert space which is in agreement with $\mathcal{H}_1(P)$.

Suppose that a compound Poisson process is given. Then, we can form the space $\mathcal{H}_1^{(-1)}(P)$ where $P'(u)$ lives. Hence, there is an infinitesimal random variable at u, tacitly assumed that $t = 1$.

In reality, we can take a space of test functionals and consider the canonical bilinear form connecting $\mathcal{H}_1^{(1)}(P)$ and $\mathcal{H}_1^{(-1)}(P)$ which defines the $P'(u)$.

The basic idea is that $P'(u)$ has identity in $\mathcal{H}_1^{(-1)}(P)$. Then we come to the method of identification of $P'(u)$ for *each* u. Recall the definition of the space $\mathcal{H}_1^{(-1)}(P)$; namely it is the dual space of $\mathcal{H}_1^{(1)}(P)$ which is separable. This means that each $P'(u)$ can be identified by at most *countably* many members in the space $\mathcal{H}_1^{(1)}(P)$;

Each member of $\mathcal{H}_1^{(1)}(P)$ is concrete and visualized. This means that $P'(u)$ can be obtained with the help of members of $\mathcal{H}_1^{(1)}(P)$; to have the canonical bilinear forms.

8.5 Nonlinear functionals of P' with special emphasis on quadratic functionals

We can also consider the space of nonlinear functionals of P'. Like $\mathcal{H}_1(P)$, as in the linear case, the spaces $\boldsymbol{F}_n$, $n \geq 2$ can be defined as a subspace of the Reproducing Kernel Hilbert Space $\boldsymbol{F}$.

Like Gaussian case, we can define spaces of generalized functionals of degree $n(\geq 2)$: $\boldsymbol{F}_n^{(-n)}$ and discuss their representations. We have

$$\boldsymbol{F}_n^{(-n)} \cong \mathcal{H}_n^{(-n)} \tag{8.5.1}$$

As soon as nonlinear functionals of the $P'(u)$ is discussed, we generally need renormalization as in the case of $\dot{B}(t)$. There is one aspect to understand the necessity of renormalization as follows.

There are many reasons why the quadratic forms, among others nonlinear functionals, are twofold.

i) If the second variation of general functionals is considered, we are given similar expressions of *normal functionals* in the Lévy's sense [85]. In terms of Reproducing Kernel Hilbert Space defined by $C^P(\xi)$ we take the expression

$$\begin{aligned}
&\int f(u)(e^{i\xi(u)u} - 1)^2 u^2 \eta(u)^2 \lambda(u)^2 du \\
&+ \iint F(u,v)(e^{i\eta(u)u} - 1)(e^{i\eta(v)v} - 1)uv \\
&\qquad \eta(u)\eta(v)\lambda(u)\lambda(v)dudv,
\end{aligned} \tag{8.5.2}$$

where f and F satisfy integrability conditions, respectively, and where $F(u,v)$ is symmetric. Such an expression comes from "passage from digital to analogue". The idea is the same as in the Gaussian case, so the details are omitted. See the monograph[57].

ii) The kernel function $F(u,v)$ plays a role of integral operator, so that we are ready to appeal to the classical functional analysis.

The most important fact is that for the first term of the equation (8.5.2), we need *renormalization* of polynomials in $P'(u)$.

For the construction of $\mathcal{H}_n^{(-n)}(P)$, we refer the monograph[57].

8.6 Observations

The characteristic functional is

$$\begin{aligned} C(\xi) &= e^{\int (e^{iu\xi(u)}-1)\lambda(u)du} \\ &= E\left(e^{i\int P'(u)\xi(u)du}\right) \\ &= \int e^{i\langle x,\xi\rangle} d\nu_\lambda^P, \end{aligned}$$

where x is a sample function of $P'(u) = P'(u,x), u \in (0,\infty)$ and ν_λ^P is defined on E_1^*.

Approximate $\lambda(u)du$ by δ-measure.

i) Let $\lambda(u) = \delta_{u_0}(u)\lambda_0$, then

$$C(\xi) = e^{\lambda_0(e^{iu_0\xi(u_0)}-1)}.$$

It is a characteristic function of Poisson type variable X_0 evaluated at $\xi(u_0)$, i.e. $E(e^{i\xi(u)X_0})$.

Thus X_0 is approximated by $u_0 P(\lambda_0)$, where $P(\lambda_0)$ is a Poisson variable with intensity λ_0.

ii) Let $\lambda(u) = \sum \delta_{u_k}(u)$, then

$$C(\xi) = \exp\left(e^{i\sum_k \lambda_k u_k \xi(u_k)} - 1\right)$$

which is actually

$$E\left(e^{i\sum_k \xi(u_k)X_k}\right),$$

where $\{X_k\}$ is independent and $X_k \approx u_k P(\lambda_k)$.

Mean of $X = \sum \xi(u_k) X_k$ is

$$\sum u_k \xi(u_k) \lambda_k$$

and its variance is

$$\sum (u_k)^2 \xi(u_k)^2 \lambda_k.$$

Take λ_u as before and take $\lambda_k = \lambda(u_k)\Delta_k$ where $\{\Delta_k\}$ is partition of $(0, \infty)$. When $\max|\Delta_k|$ tends to zero then the mean of X tends to $\int u\xi(u)\lambda(u)du$ and the variance of X tends to $\int u^2\xi(u)^2\lambda(u)du$.

Hence the limit of $\sum_k \xi(u_k)X_k$ may be thought of as an approximation of the integral

$$\int \xi(u)X(u)du$$

to be defined.

Note that the characteristic functional of the integral defined above is

$$e^{\int (e^{iu\xi(u)}-1)\lambda(u)du}$$

which is in agreement with that of $P'(u)$.

Thus we see that $P'(u)$ is realized by $X(u)$ which is an infinitesimal Poisson random variable.

The observation made above tells that the choice of the characteristic functional is reasonable and it defines a noise depending on the space (or scale), actually we have revisited the understanding of the $P'(u)$.

Chapter 9

Lévy Processes

This chapter is devoted to Lévy processes and their time derivative, i.e. the Lévy noises.

What we have discussed so far are heavily depending on analysis over Hilbert spaces and harmonic analysis arising from the transformation group, although the topics are restricted to those related to Gaussian noise and Poisson noise.

Coming back to the idea of reduction of random complex system, we may be led naturally to think of another system of elemental random variables. If we note the Lévy-Itô decomposition of a Lévy process, we can take a compound Poisson process and the noise obtained by its time derivative. It is not atomic, so it is recommended to take an elemental, i.e. atomic component, which is a Poisson noise.

We shall see much similarity between white noise, which is Gaussian, and the Lévy noise although the latter is compound. From our viewpoint, where independence is the key notion, it is natural to see similarity, indeed, two noises are generalized stochastic processes with independent values at every time t, for which the probability distributions of them are introduced on the space of generalized functions.

As soon as we come to the investigation of the properties of their probability distributions we can recognize dissimilarity. It gives good contrast, and at the same time it tells the significance of both noises.

As far as L^2-theory is concerned, (Gaussian) white noise and Poisson noise can be discussed in a similar manner, although formulas of functionals and operators present different expressions. For instance, difference can be seen in the sample function properties; Brownian motion has continuous path, while paths of Lévy process are ruled function.

As for the L^2-space, Gaussian case depends on Hermite polynomials, while functionals of a Poisson process, being a component of Lévy process, Chalier polynomials play the basic role.

We also see dissimilarity of the invariance between Gaussian measure and Poisson measure by using the transformation group acting on.

This chapter is therefore provided not as a supplementary chapter, but to see the difference of probabilistic properties among noises in which we are much involved. We now emphasize the significance of *dissimilarity* between these two noises. Significant differences appear in sample function behavior and in groups under which associated measures are kept invariant; one is the rotation group and the other is the symmetric group.

9.1 Additive process and Lévy process

We first give the definition of additive process.

Definition 9.1 A stochastic process $X(t), t \geq 0$, is called an *additive process* if it satisfies the following conditions:

1) For any $n \geq 1$ and for any t_j's with $0 \leq t_0 < t_1 < \cdots < t_n$,

$$X(t_0), X(t_1) - X(t_0), X(t_2) - X(t_1), \cdots, X(t_n) - X(t_{n-1})$$

are independent.

2) $X(0) = 0, a.e.$
3) Continuous in probability, i.e. for any $t, s \geq 0$, and for any $\epsilon > 0$

$$\lim_{s \to t} P(|X(s) - X(t)| > \epsilon) = 0.$$

The following property holds.

If $X(t)$ is an additive process, then there exists a version $X(t, \omega)$ such that $X(t, \omega)$ is right continuous and has left limit, namely $X(t, \omega)$ is a ruled function.

Definition 9.2 An additive process $X(t), t \geq 0$, is called a *Lévy process* if the probability distribution of $X(t + h) - X(t)$ is independent of t.

Example 9.1 Brownian motion and Poisson process are additive processes.

In particular, the processes which satisfy the self-similarity are interesting. Intuitively speaking, if the graph of a sample function is modified in such a way that time variable is dilated and the scale is changed according to the change of the time scale, then the graph coincides with the original one. This property can rigorously be stated as follows.

An additive process is considered to be typical if it has *stationary* independent increments. More precisely, $X(t+h)-X(t), h > 0$, the independent increment of the process $X(t), t \geq 0$, has the probability distribution which depends only on h. In particular, let $t_k = kh, h > 0, k = 0, 1, \cdots, n$, and $t_n = t$ then

$$X(t) = \sum_{k=0}^{n-1} \left(X(t_{(k+1)h}) - X(t_{kh})\right), \quad t_{nh} = t,$$

which is the sum of independent identically distributed random variables. Therefore the probability distribution of $X(t)$ is the convolution of probability distribution of $X(t_{(k+1)h}) - X(t_{kh})$, $0 \leq k \leq n-1$.

Since n is arbitrary, the distribution of $X(t)$ is a convolution of many distributions. Namely, $X(t)$ is *infinitely divisible.* (See Definition 9.3.)

Notation : We write the convolution $\mu^{n*} = \mu * \mu * \cdots * \mu$ (as many as n).

Definition 9.3 For any positive integer n, there exists μ_n such that

$$\mu = \mu_n^{n*}$$

then the probability distribution μ is *infinitely divisible.*

Let $\varphi(z)$ and $\varphi_n(z)$ be the characteristic functions associate to the probability distributions μ and μ_n respectively. Then,

$$\varphi(z) = \varphi_n(z)^n.$$

Example 9.2 Gaussian distribution $N(m, \sigma^2)$ is infinitely divisible. We can see from the following equality that holds for every n.

$$\exp\left[imz - \frac{\sigma^2 z^2}{2}\right] = \left(\exp\left[i\frac{m}{n}z - \frac{(\sigma^2/n)z^2}{2}\right]\right)^n.$$

Remark 9.1 *Note that the notion that a Gaussian distribution is infinitely divisible is different from the fact that it is an atomic random variable.*

Theorem 9.1 *(Lévy-Khintchine) The characteristic function of an infinitely divisible distribution $\varphi(z)$ can be expressed as*

$$\varphi(z) = \exp[\psi(z)]$$

and where

$$\psi(z) = imz - \frac{\sigma^2}{2}z^2 + \int_{-\infty}^{\infty}\left(e^{izu} - 1 - \frac{izu}{1+u^2}\right)dn(u), \tag{9.1.1}$$

and m is integer and

$$\int_{|u|>1} dn(u) < \infty, \int_{|u|\leq 1} u^2 dn(u) < \infty$$

and $dn(\{0\}) = 0$.

Definition 9.4 The measure dn, appeared in the above theorem, on R^1 is called the *Lévy measure.*

The expression of the characteristic function of infinitely divisible distribution is known as the standard formula due to Lévy-Khinchine. The proof of this formula is explained in Lévy[81], K.Itô, K. Sato and others, so it is omitted here.

We now remind the characteristic function of $X(t)$ which is a linear combination of independent Poisson processes. Take infinitesimal Poisson process with intensity $dn(u)$, set $t = 1$, subtract $\frac{u}{1+u^2}$ and combine them. Then, we see that it corresponds to the third term, the integral part of $\psi(z)$ expressed in (9.1.1), where the term $\frac{izu}{1+u^2}$ requires for the convergence of the integral.

The first term $\boldsymbol{imz}$ corresponds to the mean and $\boldsymbol{\frac{\sigma^2}{2}z^2}$ denotes the Gaussian distribution with mean 0 and variance σ^2.

Let us consider the probability distribution of a temporary homogeneous additive process $X(t), t \geq 0$, that has been discussed in the beginning of this section. Fix an instant t arbitrarily. Divide the interval $[0, t]$ into n subintervals $\Delta_k = [t_k, t_{k+1}]$ with same length; $t_k = \frac{kt}{n}, k = 0, 1, 2, \cdots, n$.

By assumption $\Delta_k X = X(t_{k+1}) - X(t_k), k = 0, 1, \cdots, n-1$, is a system of independent, identically distributed random variables. Denote the characteristic function of the increments $\Delta_k X$ of $X(t)$ by $\varphi_k(z)$ of their sum and $\varphi(z,t)$ respectively. Then, from the additive property of $X(t)$, we have

$$(\varphi_k(z))^n = \varphi(z,t).$$

Thus, the following proposition can be proved similarly by the additive property and stochastic continuity of $X(t)$ in t.

Proposition 9.1 *Let $X(t)$ be a Lévy process, the probability distribution of which be μ_t and the characteristic function be $\varphi(z,t)$, then*

i) μ_t is infinitely divisible for each t.

ii) μ_t can be determined by a function ψ where $\psi(z,t) = \log \varphi(z,t)$ and $\psi(0,t) = 0$. In addition,

$$\psi(z,t) = t\psi(z).$$

Since the probability distribution of Lévy process $X(t)$ is infinitely divisible, it is a composition of a Gaussian and compound Poisson processes and we can see that the probabilistic structure of the original Lévy process.

Theorem 9.2 *Let $X(t), t \geq 0$, be the Lévy process. Then $X(t)$ can be decomposed :*

$$X(t) = imt + \sigma B(t) + \lim_{p \to \infty} \int_{|u|>p} \left(uP_{du}(t) - \frac{tu}{1+u^2} \right) dn(u), \qquad (9.1.2)$$

where $B(t)$ is Brownian motion, dn is Lévy measure and P_{du} is infinitesimal Poisson random variable.

For the proof see Hida[22].

9.2 Compound Poisson process

Let $P(t)$ be a Poisson process. The characteristic function of $P(t), t \geq 0$ is

$$E(\exp[izP(t)]) = \exp[\lambda t(e^{iz} - 1)]$$

where λ is *intensity.* Sample function of $P(t)$ is a non-negative integer valued increasing step function. We may assume that it is right continuous.

The distribution of the length of the time interval, where $P(t)$ is constant, is exponential distribution with mean λ^{-1}.

From this property, a next step, we consider a process as the generalization with this property. A process $cP(t)$ with constant c is called a Poisson type process. NA compound Poisson process which is superposition of independent Poisson type processes. By this method a class of new interesting stochastic processes is introduced.

Start from Poisson processes. Note that a Poisson process can be determined by a single parameter λ which is positive and called the intensity.

Let $P_1(t)$ and $P_2(t)$ be independent Poisson processes with intensities λ_1 and λ_2, respectively and let

$$X(t) = a_1 P_1(t) + a_2 P_2(t), \ t \geq 0,$$

then there are three cases.

1) Consider the most simple case such that $a_1 = a_2 = 1$ and let the intensities of $P_1(t)$ and $P_2(t)$ be λ_1 and λ_2, respectively. That is

$$X(t) = P_1(t) + P_2(t).$$

It can be easily seen that $X(t)$ is also an additive process. The probability distribution of $X(t)$ is

$$\begin{aligned} P(X(t) = n) &= \sum_{j=0}^{n} P(X_1(t) = j, X_2(t) = n - j) \\ &= \sum_{j=0}^{n} \frac{\lambda_1^j}{j!} e^{-\lambda_1} \frac{\lambda_2^{n-j}}{(n-j)!} e^{-\lambda_2} \\ &= \frac{(\lambda_1 + \lambda_2)^n}{n!} e^{-(\lambda_1 + \lambda_2)}. \end{aligned}$$

Namely, $X(t)$ is a Poisson process with intensity $\lambda_1 + \lambda_2$.

2) a_1 and a_2 are different and the ratio $\frac{a_1}{a_2}$ irrational.

$X(t)$ is an additive process, however it is neither a Poisson nor a Poisson type process. It can be seen from the observation of its sample function. If $X(t)$ is observed, then the value is of the form; $ja_1 + ka_2$ with non-negative integers j and k. Since $\frac{a_1}{a_2}$ is irrational numbers, we know the values $P_1(t) = j$ and $P_2(t) = k$.

Since the intensities are different, asymptotic behavior (as $t \to \infty$) of sample functions discriminates the two.

In addition, consider a general form of the sum of a number of Poisson processes. It is different with the sum of Brownian motions in stochastic behavior and we can see something intrinsic.

Theorem 9.3 *Let*

$$X(t) = \sum_{1}^{\infty} a_j P_j(t), \ t \geq 0, \tag{9.2.1}$$

where $P_j(t)$'s are mutually independent Poisson processes with the same intensity λ, $a_n \geq 0$ and $\sum_1^\infty a_n < \infty$, then the series (9.2.1) is convergent almost everywhere and $X(t)$ is additive process with stationary independent increments.

Proof. The above series is the sum of random variables which are independent for each t. The n-th term has mean $a_n \lambda t$ and variance $a_n^2 \lambda t$. Since $\sum a_n \lambda t < \infty$ and $\sum a_n^2 \lambda t < \infty$, the above series is convergent almost everywhere. We also know that the sum $X(t)$ is an additive process.

■

Proposition 9.2 *Consider a finite sum*

$$X(t) = \sum_{1}^{n} a_j P_j(t), \ t \geq 0, \tag{9.2.2}$$

where $P_j(t)$'s are mutually independent Poisson processes with the same intensity. If the ratio of any two different a_j in (9.2.2) is not a rational number, the sample function of $X(t)$ can be decomposed into $P_j(t)$'s.

Since the characteristic function of Poisson process $P(t)$ with intensity λt is

$$\varphi(z) = \exp[t\lambda(e^{izt} - 1)]$$

then the characteristic function of $a_j P_j(t)$ is

$$\exp\left[t\lambda(e^{ia_j z} - 1)\right].$$

Noting $P_j(t), 1 \leq j \leq n$ are linearly independent we have the following proposition.

Proposition 9.3 *For a fixed time t the characteristic function of $X(t)$, given by (9.2.1), is*

$$\varphi_X(z) = \exp\left[t\lambda \sum_j \left(e^{iz\sum a_j} - 1\right)\right]$$

Proposition 9.4 *i) $X(t)$, given by (indep-sum) is mean continuous.*
ii) Almost all sample functions of $X(t)$ has no fixed discontinuity, but they are right continuous.

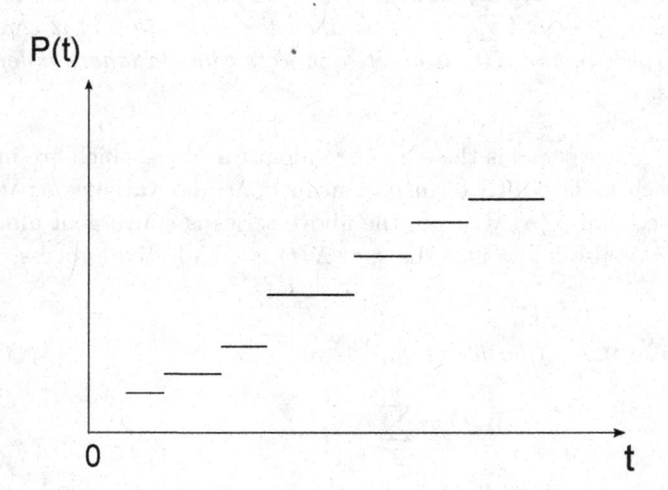

Fig. 9.1

Proof. i) Set $\sum a_j = a$. Then, $E(X(t)) = \sum a_j \lambda t = a\lambda t$.
Consequently,

$$\begin{aligned}
&E((X(t+h) - X(t))^2) \\
&= E\{(X(t+h) - X(t) - a\lambda h)^2\} - (a\lambda h)^2 + 2a\lambda h E(X(t+h) - X(t)) \\
&= \sum a_j^2 E(P_j(t+h) - P_j(t) - \lambda h)^2) + (\lambda h)^2 \\
&= a^2 \lambda h + (\lambda h)^2,
\end{aligned}$$

which converges to 0 when $h \to 0$. That is mean continuous.

ii) Taking the length h of time interval being small enough, it is impossible that more than two sample functions of $P_j(t)$ jump in a small time interval. That is, there will only be one jump with height a_j in time h or constant in the neighborhood of that time point.

■

Letting the coefficients a_j vary and letting the number of the summand increase with compensation of the summand necessary constants so as to exist reasonable sum, we have a compound Poisson process. We now have a naive, in fact intuitive construction, however the sum that we have discussed would serve a good interpretation of the general theory of the decomposition of a Lévy process.

The reader can see the idea by Fig. 9.2 in the following.

Lévy's caricature of compound Poisson process

$\varphi(t)$: periodic, period 1, $\varphi(t) = -t$, on $(-\frac{1}{2}, \frac{1}{2})$

$u_n(t) = \frac{1}{2^n}\varphi(2^n t)$,

$f(t) = \sum_0^n u_k(t)$

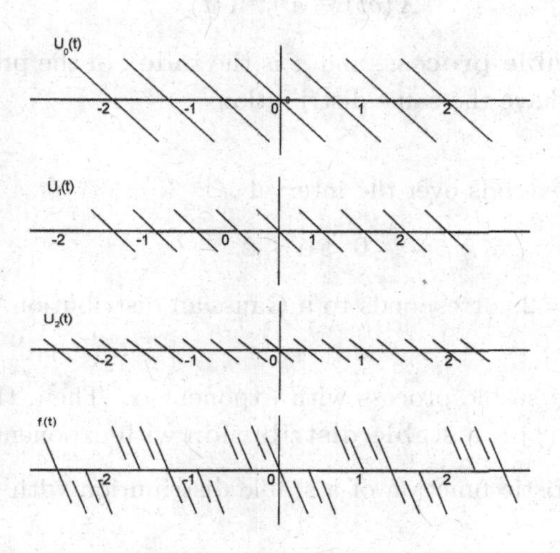

Fig. 9.2

Figure 9.2 illustrates the Lévy idea to observe the way of composition of Poisson processes from the sample function viewpoint.

Stable process which will be discussed in the next section can be said a limit of compound Poisson process, in a sense.

Discussions given so far can help to understand the meaning of the decomposition of a Lévy process stated in Theorem 9.2.

9.3 Stable distribution

We first define the stable distribution.

Definition 9.5 Let X and Y be independent random variables having the same probability distribution function $F(x)$. If for any positive constants a_1, a_2 and for any b_1, b_2, there exists a random variable Z having this distribution such that $a_1X + b_1 + a_2Y + b_2$ and $cZ + d$ with $c > 0$ have the same distribution, then F is called a **stable distribution**.

Definition 9.6 Let $X(t), t \geq 0$, be Lévy process. For any $a > 0$ and t there exists $\alpha > 0$ such that

$$X(at) \cong a^{1/\alpha} X(t),$$

then $X(t)$ is a **stable process**, and α is the **index** of the process. Here $\cong$ means that they have the same distributions.

The index α extends over the interval

$$0 < \alpha \leq 2.$$

In particular, $\alpha = 2$ corresponds to a Gaussian distribution. See Example 9.2.

Let $X(t)$ be a stable process with exponent α. Then, the probability distribution of $X(1)$ is a **stable distribution** with exponent α.

The characteristic function of a stable distribution with exponent α is of the form

$$\varphi(z) = \exp[-c|z|^{\alpha}], \quad Re\ c > 0.$$

Since the normal distribution, the Cauchy distribution, and the Lévy distribution all have the above property, it follows that they are special cases of stable distributions.

Example 9.3 A Gaussian distribution is a stable distribution with exponent 2 as is easily seen from the formula of characteristic function. (See Example 9.2.)

Example 9.4 The Cauchy distribution is a stable distribution with exponent 1.

It has two parameters $a(>0)$ and b. The probability distribution function $F_1(x)$ is of the form

$$F_1(x) = \frac{1}{2} + \frac{1}{\pi}\tan^{-1}\frac{x-b}{a}.$$

Hence, the probability density function $f_1(x)$ is

$$f_1(x) = \frac{1}{\pi}\frac{a}{a^2+(x-b)^2}.$$

Example 9.5 A stable distribution over $[0,\infty)$ with exponent $\frac{1}{2}$ has a parameter $a(>0)$ and its density function $p_{\frac{1}{2}}(x)$ is expressed in the form

$$p_{\frac{1}{2}}(x) = \frac{a}{\sqrt{2\pi}}\exp[-\frac{a^2}{2x}]x^{-3/2}, \quad x>0.$$

This distribution has appeared in Section 8.3 in connection with the inverse of maximum of a Brownian motion.

Remark 9.2 (*D. Raikov*)

Let X and Y be independent and subject to Poisson distribution (intensity may be different). Then the sum $X+Y$ is also a Poisson distribution. That is, a Poisson process has the renewal property. But it is not a stable distribution.

In order to have stable property there must have a convolution of Poisson distribution with various intensities and with suitable weight.

Remark 9.3 *There may be an idea to discuss stable distribution before stable process. We wish, however, to deal with stable distribution as a representation of a stable process at an instant, say at t. The main reason is that additive property of a stable process means that it superposed various suitable random events before we come to the instant. There should be "smearing effect" which implies smoothness of the distribution. We can think of the concept "domain of attraction".*

With this observation, we understand stable distribution naturally comes after stable process.

This observation gives us important suggestion in applications, e.g. statistical study of fractional power distributions, often called fat tail distributions that are discriminated by Gaussian distribution. If a given distribution is not Gaussian, but smooth frequency density, it is quite natural to think of the history that the random events are smeared. Interesting fact is that we can see Lévy's idea behind.

Appendix

A.1 Sobolev spaces

Sobolev space is defined as

$$K^m(R^n) = \left\{f; \hat{f}(1+\lambda^2)^{\frac{m}{2}} \in L^2\right\}, \ m > 0.$$

Thus, for $m = 0, K^0(R^n) = L^2(R^n)$.

$K^{-m}(R^n)$ is a dual space of $K^m(R^n)$ with respect to the $L^2(R^n)$-norm. That is

$$K^m(R^n) \subset L^2(R^n) \subset K^{-m}(R^n).$$

A.2 Hermite polynomials

Definition A.1 Hermite polynomial of degree n is defined by

$$H_n(x) = (-1)^n e^{x^2} \frac{d^n}{dx^n} e^{-x^2}, \quad n \geq 0. \tag{A.2.1}$$

Moment Generating Function :

$$\sum_{n=1}^{\infty} \frac{t^n}{n!} H_n(x) = e^{-t^2+2tx} \tag{A.2.2}$$

$$H_n(x) = n! \sum_{k=0}^{[\frac{n}{2}]} \frac{(-1)^k}{k!} \frac{(2x)^{n-2k}}{(n-2k)!} \tag{A.2.3}$$

Examples

$$H_0(x) = 1; \quad H_1(x) = 2x; \quad H_2(x) = 4x^2 - 2 \tag{A.2.4}$$

$$H_3(x) = 8x^3 - 12x; \quad H_4(x) = 16x^4 - 48x^2 \tag{A.2.5}$$

$$H_5(x) = 32x^5 - 160x^3 + 120x \tag{A.2.6}$$

$$H_n''(x) - 2xH_n'(x) + 2nH_n(x) = 0 \tag{A.2.7}$$

$$H_n'(x) = 2nH_{n-1}(x) \tag{A.2.8}$$

$$H_{n+1}(x) - 2xH_n(x) + 2nH_{n-1}(x) = 0 \tag{A.2.9}$$

$$H_n(ax + \sqrt{1-a^2}y) = \sum_{k=0}^{n} \binom{n}{k} a^{n-k}(1-a^2)^{k/2} H_{n-1}(x)H_k(y) \tag{A.2.10}$$

Particular Cases

$$H_n(\frac{x+y}{\sqrt{2}}) = 2^{-n/2} \sum_{k=0}^{n} \binom{n}{k} H_{n-k}(x)H_k(y) \tag{A.2.11}$$

$$H_m(x)H_n(x) = \sum_{k=0}^{m \wedge n} 2^k k! \binom{m}{k} \binom{n}{k} H_{m+n-k}(x) \tag{A.2.12}$$

$$\int_{-\infty}^{\infty} H_m(x)H_n(x)e^{-x^2}\,dx = \sqrt{\pi}\delta_{n;m}2^n n!\sqrt{\pi} \tag{A.2.13}$$

$$\int_{-\infty}^{\infty} H_m(\frac{x}{\sqrt{2}})H_n(\frac{x}{\sqrt{2}})e^{-x^2/2}dx = \sqrt{2\pi}\delta_{n;m}2^n n! \tag{A.2.14}$$

Set

$$\xi_n(x) = \frac{1}{\sqrt{2}^n n! \pi^{1/4}} H_n(x)e^{-x^2/2} : \ \{\xi_n; n \geq 0\}, \tag{A.2.15}$$

is a complete orthonormal system in $L^2(R)$.

$$\xi_n'(x) = \sqrt{\frac{n}{2}}\xi_{n-1}(x) - \sqrt{\frac{n+1}{2}}\xi_{n+1}(x) \tag{A.2.16}$$

$$x\xi_n(x) = \sqrt{\frac{n}{2}}\xi_{n-1}(x) + \sqrt{\frac{n+1}{2}}\xi_{n+1}(x) \tag{A.2.17}$$

$$(-\frac{d^2}{dx^2} + x^2 + 1)\xi_n = (2n+2)\xi_n \tag{A.2.18}$$

$$\frac{1}{\sqrt{2\pi}}\int_{-\infty}^{\infty} e^{ixy}H_n(y)e^{-y^2/2}dy = i^n H_n(x)e^{-\frac{x^2}{2}} \tag{A.2.19}$$

$$\frac{1}{\sqrt{2\pi}}\int_{-\infty}^{\infty} e^{ixy}H_n(y)e^{-y^2}dy = \frac{1}{\sqrt{2}}(ix)^n e^{-\frac{x^2}{4}} \tag{A.2.20}$$

$$\frac{1}{\sqrt{2\pi}}\int_{-\infty}^{\infty} (x+iy)^n e^{-y^2/2}dy = 2^{-n/2}H_n(\frac{x}{\sqrt{2}}) \tag{A.2.21}$$

$$\frac{1}{\sqrt{2\pi}}\int_{-\infty}^{\infty} (H_n(\frac{i}{\sqrt{2}}x + y)e^{-y^2/2}dy = i^n H_n(\frac{x}{\sqrt{2}}) \tag{A.2.22}$$

Hermite polynomials with parameter

Definition A.2 Hermite polynomials of degree n with parameter is defined as

$$H_n(x;\sigma^2) = \frac{(-\sigma^2)^n}{n!}e^{\frac{x^2}{2\sigma^2}}\frac{d^n}{dx^n}e^{-\frac{x^2}{2\sigma^2}}; \quad \sigma > 0;\ n \geq 0 \tag{A.2.23}$$

Generating function :

$$\sum_{n=1}^{\infty}\frac{t^n}{n!}H_n(x;\sigma^2) = e^{-\frac{\sigma^2}{2}t^2+tx} \tag{A.2.24}$$

$$H_n(x;\sigma^2) = \frac{\sigma^n}{n!2^{n/2}}H_n(\frac{x}{\sqrt{2}\sigma}) \tag{A.2.25}$$

$$H_n(x;\sigma^2) = \frac{\sigma^n}{2^{n/2}}\sum_{k=0}^{[n/2]}\frac{(-1)^k}{k!}\frac{(\sqrt{2}x/\sigma)^{n-2k}}{(n-2k)!} \tag{A.2.26}$$

$$H''_n(x;\sigma^2) - \frac{x}{\sigma^2}H'_n(x;\sigma^2) + \frac{n}{\sigma^2}H_n(x;\sigma^2) = 0 \tag{A.2.27}$$

$$H'_n(x;\sigma^2) = H_{n-1}(x;\sigma^2) \tag{A.2.28}$$

$$H_{n+1}(x;\sigma^2) - \frac{x}{n+1}H_n(x;\sigma^2) + \frac{\sigma^2}{n+1}H_{n-1}(x;\sigma^2) = 0 \tag{A.2.29}$$

$$\sum_{k=0}^{n} H_{n-k}(x;\sigma^2)H_k(y;\tau^2) = H_n(x+y;\sigma^2+\tau^2) \tag{A.2.30}$$

$$H_m(x;\sigma^2)H_n(x;\sigma^2) = \sum_{k=0}^{m\wedge n} \frac{\sigma^{2k}(m+n-2k)!}{k!(m-k)!(n-k)!}H_{m+n-2k}(x;\sigma^2) \tag{A.2.31}$$

Set $\eta_n(x;\sigma^2) = \frac{\sqrt{n!}\sigma^n}{H}{}_n(x;\sigma^2)$, $\{\eta_n; n \geq 0\}$ is a complete orthonormal Gaussian system in $L^2(R, \frac{1}{\sqrt{2\pi}\sigma}e^{-\frac{x^2}{2\sigma^2}}dx)$.

A.3 Rademacher functions

On the measure space $([0,1].\mathcal{B}, dx)$, Rademacher functions $r_n(x), n > 1$ are defined as follows.

Let the binary expansion of $x \in [0,1]$ be

$$x = 0.\varepsilon_1\varepsilon_2\cdots.$$

Let $\varepsilon_n = \varepsilon_n(x)$.

Set

$$r_n(x) = \begin{cases} 1, & \varepsilon_n(x) = 1, \\ -1, & \varepsilon_n(x) = 0, \\ 0, & \text{if } x \text{ is expandable in two ways.} \end{cases}$$

The functions $r_n(x), n > 1$ are called the Rademacher functions. They form a sequence of orthonormal system, but they are not complete.

A.4 Spectral decomposition of covariance function

Let $X(\varphi), \varphi \in E$ be any stationary random distribution with covariance functional $\rho(\varphi)$.

Then the spectral decomposition of ρ is

$$\rho(\phi) = \int \hat{\phi}(\lambda) d\mu(\lambda),$$

where $\hat{\phi}$ is the Fourier transform of ϕ.

K. Itô classified the stationary random distribution as follows.
If μ satisfies

$$\int \frac{d\mu(\lambda)}{(1+\lambda^2)^k} < \infty$$

then X is in $\boldsymbol{\mathcal{S}}_k$.

Let $\boldsymbol{\mathcal{S}}$ be the totality of stationary random distributions.
Then

$$\boldsymbol{\mathcal{S}} = \cup_k \boldsymbol{\mathcal{S}}_k,$$

$$\cdots \subseteq \boldsymbol{\mathcal{S}}_{-2} \subseteq \boldsymbol{\mathcal{S}}_{-1} \subseteq \boldsymbol{\mathcal{S}}_0 \subseteq \boldsymbol{\mathcal{S}}_1 \subseteq \boldsymbol{\mathcal{S}}_2 \subseteq \cdots.$$

A.5 Variational calculus for random fields

This section is devoted to the variational calculus of *random fields.* This topic is one of the most important area where the white noise theory is efficiently applied.

The main subjects of random field are

1) First we take stochastic processes with multi-dimensional parameter. The typical and indeed most important example is Lévy's Brownian motion. It is a Gaussian system that has most interesting way of dependence. So, we shall discuss in the next section.

2) The second is a system of random variables indexed by a manifold in the higher dimensional Euclidean space. Let $\mathbf{C}$ be a class of smooth, closed and convex manifolds C with co-dimension 1, in the d-dimensional

Euclidean space R^d. The topology that is introduced to $\mathbf{C}$ is naturally induced by the Euclidean distance.

We are particularly interested in the Gaussian case, that is, the collection $\mathbf{X} = \{X(C), C \in \mathbf{C}\}$ is a Gaussian system.

We shall mainly deal with Gaussian random fields, since our study of them heavily depends on the white noise analysis. The results are therefore satisfactory in a sense.

We would like to explain here that the idea of the study has come from the innovation theory in these cases 1) and 2).

The innovation theory actually depends on the Lévy's idea which has been stated in his 1953 Univ. of Calif, Berkeley Publication. He states that a structure of a stochastic process $X(t)$ is characterized by the equation

$$\delta X(t) = \varphi(X(s), s \leq t, Y(t), t, dt), \tag{A.5.1}$$

which is called the *infinitesimal equation.* The variable $Y(t)$ is an infinitesimal random variable that is independent of $X(s), s \leq t$, and contains full information that the process $X(t)$ gains during the infinitesimal time interval $(t, t + dt]$. We call $Y(t)$ the *innovation* of the process $X(t)$.

Of course, the above equation is quite formal, however it tells us the general idea of investigating probabilistic structure of a stochastic process.

It is our problem how to concretize the infinitesimal equation. For Gaussian systems we can, in a sense, realize the idea.

For example

a) A Gaussian process. There is an established theory on canonical representation of a Gaussian process. We need, of course, some essential assumptions, like unit multiplicity and no remote past.

   With those assumptions, a given Gaussian process $X(t)$ is expressed in the form

$$X(t) = \int^t F(t,u)\dot{B}(u)du,$$

   where $F(t, u)$ is a sure function. The significant part is that canonical property of the representation asserts that the $\dot{B}(t)$ is the innovation. Thus the problem raised by infinitesimal equation is solved.

Very particular and interesting case is found in the theory of strictly multiple Markov Gaussian process $X(t)$. The infinitesimal equation is specified to be an ordinary differential equation involving time derivatives $X^{(k)}(t)$ and $\dot{B}(t)$ which is rigorously defined in white noise theory as a member of the space $\mathcal{H}_1^{(-1)}$. The given equation is

$$\sum_0^N a_k(t) X^{(N-k)}(t) = \dot{B}(t), \quad 0 \leq t \leq T.$$

With an initial condition this equation is solved. To be interested, we can form directly the canonical representation where the $\dot{B}(t)$ is the innovation. Further we can explicitly form the dual process $X^*(t), 0 \leq t \leq T$ of the given $X(t)$. These analyses can be carried on within the space of linear Hida distributions.

b) For a Gaussian random field $X(a), a \in R^d$, with $d > 1$, we hoped to have a generalization of a). It is recognized that a generalization is not in a straight manner if we are in line with white noise theory. We shall show an approach in the next section.

c) As for the case where the parameter of a random field is taken to be a manifold C like $X(C)$, we meet more difficulty. We have therefore asked the classical functional analysis for help. Some results will be presented in what follows.

Motivations of the analysis of $X(C)$

1) A study of a random field $X(a), a \in R^d$ with $d > 1$ under the viewpoint of the infinitesimal equation is not easy. If $X(a)$ is specified to be Lévy's Brownian motion, there is a famous approach by H. P. McKean[95]. We have had somewhat different approach and it will briefly be shown in the next section.

2) Random functionals, denoted by $X(C)$, depending on a manifold C running through Euclidean space R^d is called a *random field*. There are several reasons why we are interested in random fields; among others:

i) From quantum dynamics. There are some historical notes. We have been motivated by the famous P.A.M. Dirac's paper[12], where he discussed the action principle which has given a good suggestion to path integral. Thus, we should consider random fields depending on space-time variables.

ii) From classical functional analysis. We shall also be back to the classical functional analysis. We are stimulated by L. Tonelli's work on functionals, the variable of which is a manifold running through a space. See the literature[130] in his Collected papers, vol.2 1960. We can see variational equations for functionals. Then, we learn a lot like the Hadamard equation, from his theory of functionals. We then come to the story of V. Volterra's success and his own great development (ref. Volterra[131]). Later P. Lévy[85] has discussed systematically. His books are good literatures up till now.

iii) Needless to say on the Tomonaga-Schwinger equation. It is difficult to follow some of their ideas, but we are much encouraged by their works.

iv) Recent development on quantum field theory in quantum dynamics is marvelous but difficult to follow, however we are inspired.

Bibliography

1. L. Accardi, Quantum information, quantum communication and information. Quantum Information and Complexity, ed. T. Hida et al. World Scientific Pub. Co., 2004, 1-60.
2. L. Accardi and A. Boukas, White noise calculus and stochastic calculus.Stochastic Analysis : Classical and Quantum, Perspectives of White Noise Theory. ed. T. Hida, World Scientific Pub. Co., 2005, 260-300.
3. L. Accardi. T. Hida and Win Win Htay, Boson Fock Representations of Stochastic Processes, Mathematical Notes 67 (2000) 3-14
4. L. Accardi, Y. G. Lu and I. Volovich, Quantum theory and its stochastic limit. Springer-Verlag, 2002.
5. L. Accardi et al eds, Selected Papers of Takeyuki Hida. World Scientific Pub. Co., 2001.
6. L. Accardi, T. Hida and Si Si, Innovation approach to some stochastic processes. Volterra Center Pub. N.537, 2002.
7. N. U. Ahmed, Introduction to generalized functionals of Brownian motion, (to appear).
8. S. Albeverio and M. W. Yoshida, Multiple stochastic integerals construction of non-Gaussian reflection positive generalized random fields. Preprint 2006.
9. J. Berunoulli, Ars Conjectandi, 1713.
10. S. Bochner, Harmonic analysis and the theory of probability. Univ. of California Press, 1955.
11. P. A. M. Dirac, The principles of quantum mechanics. Oxford Univ. Press, 1930.
12. P. A. M. Dirac, Lagrangian in quantum mechanics. Phys. Zeitschrift Sowjetunion. 3 (1933), 64-72.
13. J. L. Doob, Stochastic processes. J. Wiley & Sons, 1952.
14. A. Einstein, Einstein's 4 papers appeared in Annalen der Physik. 17 (1905), 132-148, 549-560, 891-921; 18 (1905), 639-641.
15. M. N. Feller, The Lévy Laplacian. Cambridge Tract in Math. 166, Cambridge Univ. Press, 2005.
16. W. Feller, An introduction to probability theory and its applications. Vol I. 1950; Vol II. 1966, Wiley.

17. I .M. Gel'fand, Generalized random processes. (in Russian), Doklady Acad. Nauk SSSR, 100 (1955), 853-856.
18. I. M. Gel'fand and N. Ya. Vilenkin, Generalized functions, vol.4. Academic Press, 1964. (Russian Original, Moscow, 1961).
19. J. Glimm and A. Jaffe, Quantum physics. A functional integral point of view. Springer-Verlag, 1981.
20. M. Grothaus, D. C. Khandekar, J. L. da Silva and L. Streit, The Feynman integral for time-independent anharmonic oscillators. J. Math. Phys. 36, 1997 (3276-3299)
21. T. Hida, Canonical representations of Gaussian processes and their applications. Memoires Coll. of Sci. Univ. of Kyoto, A33 (1960), 109-155.
22. T. Hida, Stationary stochastic processes. Mathematical Notes. Princeton Univ. Press, 1970.
23. T. Hida, Note on the infinite dimensional Laplacian operator. Nagoya Math. J. 38 (1970), K. Ono volume, 13-19.
24. T. Hida, J. Multivariate Analysis, no.1 (1971) 58-69.
25. T. Hida, Complex white noise and infinite dimensional unitary group. Lecture Notes in Math. Mathematics, Nagoya Univ., 1971.
26. T. Hida, A role of Fourier transform in the theory of infinite dimensional unitary group. J. Math. Kyoto Univ. 13 (1973), 203-212.
27. T. Hida, Brownian motion. Springer-Verlag, 1980. Japanese Original: Buraun Unidou. Iwanami Pub. Co., 1975.
28. T. Hida, Analysis of Brownian functionals. Carleton Math. Notes no.13, 1975.
29. T. Hida, A note on stochastic variational equations. Exploring Stochastic Laws. Korolyuk volume. eds. A.V. Skorohod and Yu. V. Vorovskikh, 1995, 147-152.
30. T. Hida, Causal analysis in terms of white noise. Quantum Fields-Algebra, Processes. ed. L. Streit, Springer-Verlag, 1980, 1-19.
31. T. Hida, Stochastic variational calculus. Stochastic Partial Differential Equations and Applications. eds. B.L. Rozovskii and R.B. Sowers. Springer-Verlag. Lecture Notes in Math. 1992, 123-134.
32. T. Hida, A role of the Lévy Laplacian in the causal calculus of generalized white noise functionals. Stochastic Processes. G. Kallianur volume, ed. S. Cambanis, 1993, 131-139.
33. T. Hida, Random fields and quantum dynamics, Foundation of Physics. 29 (1997) Namiki volume. 1511-1518.
34. T. Hida, Some methods of computation in white noise analysis. Unconventional Models of Computation UMC'2K. eds. I. Antoniou et al. 2001, 85-93.
35. T. Hida, Innovation approach to stochastic processes and quantum dynamics. Foundations of Probability and Physics. 8 (2001), 161-169.
36. T. Hida, Complexity and irreversibility in stochastic analysis. Chaos, Soliton and Fractals 12 (2001), 2859-2863.
37. T. Hida, White noise and functional analysis.(in Japanese) Seminar on Probability Vol. 60, 2002.
38. T. Hida, White noise analysis: A new frontier. Volterra Center Pub. N.499, January 2002.

39. T. Hida, White noise analysis: Part I. Theory in Progress. Taiwanese J. of Math. 7 (2003) 541-556.
40. T. Hida, Laplacians in white noise analysis. Contemporary Mathematics. 317 (2003), 137-142.
41. T. Hida, Some of future directions of white noise analysis. Proc. Abstract and Applied Analysis. ed. N.M. Chuong (2003).
42. T. Hida, A frontier of white noise analysis in line with Itô calculus. Advanced Studies in Pure Math. 41 (2004), 111-119.
43. T. Hida, Stochastic variational equations in white noise analysis. Proc. SPDE eds. G. Da Prato and T. Chapman, Chapman & Hall/CRC, (2005) 169-178.
44. T. Hida, Information, innovation and elemental random field Quantum Information and Computing, QP-PQ, eds. L. Accardi et al. vol.19 (2006), 195-203.
45. T. Hida and M. Hitsuda, Gaussian processes. Transl. Math. Monographs 120, American Math. Soc., 1993.
46. T. Hida and N. Ikeda, Note on linear processes. J. Math. Kyoto Univ. 1 (1961), 75-86.
47. T. Hida and N. Ikeda, Analysis on Hilbert space with reproducing kernel arising from multiple Wiener integral. Proc. 5th Berkeley Sump. on Math. Stat. Prob. 2 (1967), 211-216.
48. T. Hida, I. Kubo, H. Nomoto and H. Yoshizawa, On projective invariance of Brownian motion. Pub. RIMS, Kyoto Univ. A.4, (1969), 595-609.
49. T. Hida and H. Nomoto, Projective limit of spheres. Proc. Japan Academy.
50. T. Hida, H.-H. Kuo, J. Potthoff and L. Streit, White noise. An infinite dimensional calculus. Kluwer Academic Pub. Co., 1993.
51. T. Hida, N. Obata and K. Saito, Infinite dimensional calculus and Laplacians in terms of white noise calculus. Nagoya Math. J. 128 (1992), 65-93.
52. T. Hida, J. Potthoff and L. Streit, Dirichlet forms and white noise analysis. comm. Math. Phys. 116 (1988), 235-245.
53. T. Hida and Si Si, Stochastic variational equations and innovations for random fields. Infinite Dimensional Harmonic Analysis, Transactions of a German-Japanese Symposium. eds. H. Heyer and T. Hirai, 1995, 86-93.
54. T. Hida and Si Si, Innovation for random fields. Infinite Dimensional Analysis, Quantum Probability and Related Topics. 1 (1998), 499-509.
55. T. Hida and Si Si, Innovation Approach to some problems in quantum dynamics. Garden of Quanta, Essays in Honour of Hirashi Ezawa, World Scientific Pub. Co., 2003, 95-103.
56. T. Hida and Si Si, Innovation approach to random fields: An application of white noise theory. World Scientific Pub. Co., 2004.
57. T. Hida and Si Si, Lectures on White Noise Functionals. World Scientific Pub. Co., 2008.
58. T. Hida and L. Streit, Generalized Brownian functionals. Proc. VI International Conference on Math. Physics. Berlin 1981, LN in Phys. 153 (1982), 285-287.
59. T. Hida, Si Si and Win Win Htay, A noise of new type and its application,

Ricerche di Matematica, Vol. 55, No.1, Ricerche mat.DOI 10.1007/s11587-011-0114-0.
60. M. Hitsuda, Multiplicity of some classes of Gaussian processes. Nagoya Math. J. 52 (1973), 39-46.
61. M. Hitsuda, Formula for Brownian partial derivatives, Proc. 2nd Japan-USSR Symp. Probab. Th. 2, (1972) 111-114.
62. E. Hopf, Statistical hydrodynamics and functional calculus. J. Rational Mechanics and Analysis. 1 (1952), 87-123.
63. K. Hoffman, Banach spaces of analytic functions. Prentice-Hall Inc., 1962.
64. K.Itô, On stochastic differential equations. Mem. Amer. Math. Soc. 4 (1951), 1-57.
65. K. Itô, Stationary random distributions. Mem. Univ. of Kyoto, A. 28 Math. (1953), 209-223.
66. K. Itô, Multiple Wiener integrals, J. Math. Soc. Japan 3 (1951), 157-169.
67. K. Itô and McKean Jr., Diffusion processes and their sample paths. Springer-Verlag, 1965.
68. Y. Ito and I. Kubo, Calculus on Gaussian and Poisson noises. Nagoya. Math. J. 111 (1988), 41-84.
69. G. Kallianpur, Some ramification of Wiener's ideas on nonlinear prediction. N. Wiener: Collected Works vol. III, ed P. Masani, The MIT Press, 1981, 402-424.
70. K. Karhunen, Über die Struktur stationärer zufälliger Funktionen. Arkiv för Mathematik, 1, no.13 (1950), 141-160.
71. A. N. Kolmogorov, Wienersche Spiralen und einige andere interessante Kurven in Hilbertschen Raum. Doklady Acad. Sci. URSS 26 (1940) 115-118.
72. I. Kubo and S. Takenaka, Calculus on Gaussian white noise. I - IV. Proc. Japan Academy 56 (1980), 376-380, 411-416; 57 (1981), 433-437; 58 (1982), 186-189.
73. H.-H. Kuo, Gaussian measures in Banoch space. Lecture Notes in Mathematics, 463, Springer-Verlag, 1975.
74. H.-H. Kuo, White Noise Distribution Theory, CRC Press, Probability and Stochastic Series, 1996.
75. R. Léandre, Theory of distribution in the sense of Connes-Hida and Feynman path integral on a manifold. Infinite Dimensional Analysis, Quantum Probability and Related Topics 6 (2003) 505-517.
76. R. Léandre and H. Ouerdiane, Connes-Hida calculus and Bismut-Quillen superconnections. Prépub. Institut Élie Cartan. 2003/no. 17.
77. Y. J. Lee, Analysis of generalized Lévy white noise functiomnals. J. Functional Analysis. 211 (2004), 1-70.
78. P. Lévy, Sur le équations intégro-différentielles définissant des fonctions de lignes. Thëses. 1911, 1-120.
79. P. Lévy, Lecons d'analyse fonctionelle. Gauthier-Villars, 1922.
80. P. Lévy, Sur les intégrales dont les éléments sont des variables aléatoires indépendantes, Annali Scuola norm. Pisa, s. 2,t. 3, 1934, 337-366.
81. P. Lévy, Théorie de l'addition des variables aléatoires. Gauthier-Villars. 1937. 2ème ed. 1954.

82. P. Lévy, Sur certains processus stochastiques homogènes. Compositio Mathematica Nordhoff-Groningen 7 (1939) 283-339.
83. P. Lévy, Integrales stochastiques, Societé math de France Sect. Sud-Est (1941) 67-74.
84. P. Lévy, Processus stochastiques et mouvement brownien. Gauthier-Villars, 1948. 2ème ed. 1965.
85. P. Lévy, Problèmes concrets d'analyse foctionnelle. Gauthier-Villars, 1951.
86. P. Lévy, Random functions: General theory with special reference to Laplacian random functions. Univ. of California Pub. in Statistics. 1 (1953), 331-388.
87. P. Lévy, Le mouvement Brownien, Mém. Sc. Math CXXVI, Gauthier-Villars, 1954.
88. P. Lévy, A special problem of Brownian motion, and a general theory of Gaussian random functions. Proceedings of the 3rd Bekeley Symposium on Mathematical Statistics and Probability. vol.II, Univ. of Californian Press, (1956) 133-175.
89. P. Lévy, Fonction aléatoire à corrélation linéaires. Illinois J. Math. 1, no.2 (1957), 217-258.
90. P. Lévy, Quelques aspects de la pensée d'un mathématician. Blanchard, 1970.
91. J. L. Lions and E. Magenes, Non-homogeneous boundary value problems and applications. vol.1. Springer-Verlag, 1972.
92. B. Mandelbrot and J. Van Ness, Fractional Brownian motions, fractional noises and applications. SIAM Rev. 10 (1968), 422-437.
93. P. Masani and N. Wiener, Nonlinear prediction. H. Cramér volume. John Wiley & Sons, 1959, 190-212.
94. M. Masujima, Path integrals and stochastic processes in theoretical physics. Feshback Pub. LLC. 2007.
95. H. P. McKean, Brownian motion with a several-dimensional time. Theory of Prob. and Appl. 8 (1963), 335-354.
96. J. Mikusiński, On the space of the Dirac Delta-function. Bull. de l'Acad. Polanaise des Sci. Ser. Sci. Math. Astr. Phys. 14. (1966), 511-513.
97. A.S. Monin and A.M. Yaglom, Statistical hydrodynamics. vol. 1, Nauka. 1965; vol. 2, Nauka. 1967 (in Russian).
98. N. Obata, White noise calculus and Fock space. LNM no.1577, Springer-Verlag, 1994.
99. J. Potthoff and L. Streit, A characterization of Hida distribution. J. Functional Analysis. 101 (1991), 212-229.
100. V. A. Rohlin, On the fundamental ideas of measure theory. Math. Sbornik 25 (1949), 107-150, AMS Eng. Trans Ser 1. Vol 10. (1962), 1-54.
101. K. Saitô and A. Tsoi, Stochastic processes generated by functions of the Lévy Laplacian. Quantum Information II. eds. T. Hida and K. Saitõ. World Scientific Pub. Co., 2000, 183-194.
102. C. E. Shannon and W. Weaver, The mathematical theory of communication. Univ. of Illinois Press, 1949.
103. L. Schwartz, Théorie des distributions, Hermann, 1950.

104. T. Shimizu and Si Si, Professor Takeyuki Hida's Mathematical Notes, Vol.I. 2004. (informal publication).
105. Si Si, A note on Lévy's Brownian motion I, II, Nagoya Math. J. 108 (1987) 121-130; 114 (1989), 165-172.
106. Si Si, Variational calculus for Lévy 's Brownian motion. Gaussian random fields, Series of Probability and Statistics, Satellite Conference for International Math. Congress, World Scientific Pub. Co., 1990, 364-373.
107. Si Si, Variational calculus for Lévy's Brownian motion. "Gaussian random fields" Ser. Probability and Statistics 1 (1991), 364-373.
108. Si Si, Integrability condition for stochastic variational equation. Volterra center Pub. N.217, 1995.
109. Si Si, Innovation of some random fields. J. Korea Math. Soc. 35 (1998), 575-581.
110. Si Si, A variational formula for some random fields; an analogy of Ito's formula. Infinite Dimensional Analysis, Quantum Probability and Related Topics, 2 (1999), 305-313.
111. Si Si, Random fields and multiple Markov properties, Supplementary Papers for the 2nd International Conference on Unconventional Models of Computation, UMC'2K, ed. I. Antoniou, CDMTCS-147CCenter for Discrete Mathematics and Theoretical Computer Science, Solvay Institute, 2000, pp. 64-70.
112. Si Si, Gaussian Processes and Gaussian random fields. Quantum Information U, Proc. of International Conference on Quantum Information, World Scientific Pub. Co., 2000, 195-204.
113. Si Si, X-ray data processing - An example of random communication systems, Nov. 2001.
114. Si Si, Effective determination of Poisson noise. Infinite Dimensional Analysis, Quantum Probability and Related Topics, 6 (2003), 609-617.
115. Si Si, Poisson noise, infinite symmetric group and a stochastic integral based on quadratic Hida distributions. Preprint.
116. Si Si, A characterization of Poisson noise, Quantum Probability and Infinite Dimensional Analysis, QP-PQ Vol 20, World Scientific Pub., 2007, 356-364.
117. Si Si, An aspect of quadratic Hida distributions in the realization of a duality between Gaussian and Poisson noises. Infinite Dimensional Analysis, Quantum probability and Related Topics, Vol. 11 (2008) 109-118.
118. Si Si, Multiple Markov generalized Gaussian processes and their dualities, Infinite Dimensional Analysis, Quantum probability and Related Topics, Vol. 13, (2010) 99-110.
119. Si Si, Win Win Htay and L. Accardi, T-transform of Hida distribution and factorization, Volterra Center Pub. N.625, 2008, 1-14.
120. Si Si, A. Tsoi and Win Win Htay, Invariance of Poisson noise. Proceedings: Stochastic Analysis: Classical and Quantum. ed. T. Hida, World Scientific Pub. Co., 2005, 199-210.
121. Si Si, Note on fractional power distributions (preprint).
122. Si Si and Win Win Htay, Entropy in subordination and Filtering. Acta Applicandae Mathematicae (2000) Vol. 63 No. 1-3, 433-439

123. Si Si and Win Win Htay, Structure of linear processes. Quantum Information and and computing. eds. L. Accardi et al, 2006, 304-312.
124. Si Si, A. Tsoi and Win Win Htay, Jump finding of a stable process. Quantum Information V, World Scientific Pub. Co., 2006, 193-202.
125. B. Simon, Functional integration and quantum physics. 2nd ed. AMS Chelsea Pub., 2005.
126. F. Smithies, Integral equations. Cambridge Univ. Press, 1958.
127. E.M. Stein, Topics in harmonic analysis. Related to the Littlewood-Palay theory. Ann. Math. Studies. 63, Princeton Univ. Press, 1970.
128. L. Streit and T. Hida, Generalized Brownian functional and Feynman integral. Stochastic Processes and their Applications. 16 (1983), 55-69.
129. S. Tomonaga, On a relativistically invariant formulation of the quantum theory of wave fields. Progress of Theoretical Physics. Vol.1 (1946), 27-39.
130. L. Tonelli, Fondamenti di caleolo delle variazioni, Bologna Nicola Zanichelli Editore. Vol. 1. 1921; Vol 2, 1923.
131. V. Volterra, Theory of functionals and of integral and integro-differential equations, Dover, 1959.
132. N. Wiener, Generalized harmonic analysis. Acta Math. 55 (1930), 117-258.
133. N. Wiener, The homogeneous chaos. Amer. J. Math. 60 (1938), 897-936.
134. N. Wiener, Extrapolation, interpolation and smoothing of stationary time series. The MIT Press, 1949.
135. N. Wiener, Nonlinear problems in random theory, The MIT Press, 1958.
136. Win Win Htay, Optimalities for random functions: Lee-Wiener's network and non-canonical representation of stationary Gaussian processes,Nagoya Math. J. Vol. 149 (1998), 9-17.
137. Win Win Htay, Multiple Markov Property and Entropy of a Gaussian Process (preprint)
138. Win Win Htay, Markov property and information loss. preprint. Nagoya Univ.
139. H. Yoshizawa, Rotation group of Hilbert space and its application to Brownian motion. Proceedings of International Conference on Functional Analysis and Related Topics. Tokyo. 414-423.
140. K. Yosida, Functional analysis. Springer-Verlag, 6th ed., 1980.

Index